The Comparative Method in Evolutionary Biology

Oxford Series in Ecology and Evolution
Edited by Robert M. May and Paul H. Harvey

1. The Comparative Method in Evolutionary Biology
Paul H. Harvey and Mark D. Pagel

The Comparative Method in Evolutionary Biology

PAUL H. HARVEY
Department of Zoology, University of Oxford

MARK D. PAGEL
Peabody Museum, Harvard University

Oxford New York Tokyo
OXFORD UNIVERSITY PRESS
1991

Oxford University Press, Walton Street, Oxford OX2 6DP
Oxford New York Toronto
Delhi Bombay Calcutta Madras Karachi
Petaling Jaya Singapore Hong Kong Tokyo
Nairobi Dar es Salaam Cape Town
Melbourne Auckland
and associated companies in
Berlin Ibadan

Published in the United States
by Oxford University Press, New York

British Library Cataloguing in Publication Data
Harvey, Paul H. (Paul Harvey) 1947–
The comparative method in evolutionary biology.
1. Organisms. Evolution
I. Title II. Pagel, Mark D.
575
ISBN 0–19–854641–6
ISBN 0–19–854640–8 (pbk)

Library of Congress Cataloging in Publication Data
Harvey, Paul H.
The comparative method in evolutionary biology / Paul H. Harvey and Mark D. Pagel.
p. cm.— (Oxford series in ecology and evolution: 1)
Includes bibliographical references (p.) and index.
1. Evolution. 2. Evolution-Research-Methodology. I. Pagel, Mark D. II. Title. III. Series.
QH366.2.H385 1991 575'.0072—dc20 90–21148
ISBN 0·19–854641–6
ISBN 0–19–854640–8 (pbk)

Typeset by Pentacor PLC, High Wycombe, Bucks.
Printed in Great Britain by
Courier International Ltd., Tiptree, Essex

Preface

Richard Lewontin once wrote 'That is the one point which I think all evolutionary biologists are agreed upon, that it is virtually impossible to do a better job than an organism is doing in its own environment'. What is more, many distantly related organisms work in similar environments and have often evolved similar adaptations to help them in their tasks. Viewed in this way, there are repeated patterns of evolutionary change. If evolutionary biologists think that some trait of interest may have evolved to do a particular job, they naturally use the comparative method when they ask if other organisms doing the same job have evolved similar traits.

The comparative approach is not new. Indeed it was Darwin's favoured technique. And, since his day, it has been used sooner or later by almost all right-thinking evolutionary biologists. In short, comparative studies have taught us most of what we know about adaptation. But, in little more than a decade, the comparative method has become transformed. As more data have accumulated to allow ever more finely grained comparisons, so the assumptions behind comparative tests have been made increasingly explicit, resulting in a new rigour being routinely applied to comparative studies.

Comparisons have increasingly been set against particular models of evolutionary change. It is now common to ask how we would expect particular groups of species to differ. For example, we must consider phylogenetic relatedness: closely related species will have genes in common through descent from common ancestors and are likely to be more similar in both phenotype and lifestyle than distant relatives. In other words, close relatives are likely to share the trait of interest, and the chances are that they will share a whole lot more besides! Comparative trends must, therefore, be examined together with phylogenetic relatedness. But how should this be done?

Several different methods for analysing comparative data have been suggested over the past few years. We shall review them, but we shall also attempt to achieve something more. The reasons for writing this book were to place comparative studies firmly in their biological context, to place comparative methods equally firmly in their statistical context, and to demonstrate how the two contexts must develop hand in hand. In short, all useful comparative methods are based on explicit models of evolutionary change. Although we have our own ideas about suitable evolutionary models, we have aimed to provide a framework that will be useful for comparative biologists with a variety of different views on the way evolution has proceeded. Most comparative studies may be inspired by a search for adaptive patterns but, as we shall show, many evolutionary regularities require alternative explanations.

We want to thank the following colleagues and friends who have all contributed in their different ways to the production of this book: Tim Clutton-Brock, Annie Collie, Allen Edwards, Joe Felsenstein, Ted Garland, Anna Harvey, Ben Harvey, Joe Harvey, Pam Hofer, Ray Huey, Anne Keymer, Wayne Maddison, Robert May, John Maynard Smith, Sean Nee, Andy Purvis, Andrew Read, Sir Richard Southwood, and Mark Williamson. In addition, we are grateful for the opportunity we were given to improve the first draft of this book in the light of constructive criticisms we received from Bernie Crespi, Joe Felsenstein, Ted Garland, Charles Godfray, Deborah Gordon, Peter Holland, Ray Huey, Jeremy John, Anne Keymer, Wayne Maddison, Robert May, Judith May, Sean Nee, Ted Oakes, Naomi Pierce, Daniel Promislow, Andrew Read, and Bernt-Erik Sæther. We also thank the Commission for the European Communities for supporting this work.

Finally, we must single out Joe Felsenstein for a special tribute—he alone laid the foundations on which a more rigorous comparative method could be constructed.

Oxford P.H.H.
August 1990 M.D.P.

Contents

1

The comparative method for studying adaptation

'we must learn to treat comparative data with the same respect as we would treat experimental results' (Maynard Smith and Holliday 1979, p. vii)

1.1 Introduction

It is second nature for evolutionary biologists to think comparatively because comparisons establish the generality of evolutionary phenomena. How much molecular evolution is neutral? Do large genomes slow down development? Is sperm competition important in the evolution of animal mating systems? What lifestyles select for large brains? Are extinction rates related to body size? These are all questions for the comparative method, and this book is about how such questions can be answered.

Evolutionary biology shares with astronomy and geology the task of interpreting phenomena that cannot be understood today without understanding their past. Stars in the Milky Way, mountains in the Swiss Alps, and finches in the Galápagos Islands each have their own common histories which give them characteristics that set them apart from the stars of other galaxies, mountains in other regions, and the finches of other archipelagos. Much of this book will be devoted to understanding the influence of shared phylogenetic history on the form of contemporary species. But there is something special about organisms such as finches, orchids, and aardvarks which distinguish them from the inorganic world: they have become adapted to their environments through natural selection, a process that gives life to the comparative method in evolutionary biology.

Indeed, organisms are so well adapted that a large part of organismic biology over the centuries has been devoted to the study of adaptation. Before Darwin, it was often argued that a proper understanding of adaptation might give insight into the mind of the creator. Adaptations were thought to be design features, although whether lack of adaptation was to be viewed as 'a paucity of imagination on the part of the creator' (Maynard Smith 1978, p. 136) was not so regularly posed as a serious question. Since Darwin, it has generally been accepted that adaptations have been honed by natural selection. Whichever force was to be praised or blamed, it has been a fact of nature long appreciated that different species often exhibit similar characteristics when they live in similar environments.

If similar characters evolve repeatedly in similar environments, it is reasonable to consider how they might enable their bearers to survive and reproduce in those environments. For example, several species of birds and mammals have evolved white feathers and fur in snowy environments, whereas their close relatives retain what we assume to be the ancestral plain or mottled brown coloration. Presumably, white variants were favoured by natural selection because they were cryptic against a white background of snow. For some species, such as the snowy owl (*Nyctea scandiaca*) or polar bear (*Ursus arctos*), this may have made them more effective predators, although for others, like the ptarmigan (*Lagopus mutus*) or snowshoe hare (*Lepus americanus*), a white coat probably helped to protect them from predators.

From Darwin's time to the present, the comparative method has remained the most general technique for asking questions about common patterns of evolutionary change. The comparative method has, however, changed radically in recent years, and this book is about a new type of comparative study. The major advance has been the development of methods based on explicit evolutionary and statistical models. These techniques take careful stock of the phylogenetic links between species, and marry ideas about evolutionary change with statistical processes in such a way that formal tests of hypotheses about evolution are possible. We describe these new techniques, and how to use them to study evolution and adaptation.

The motivation for many comparative studies is the occurrence of astonishing regularities that require explanation and suggest further ideas. Consider Bonner's (1965) plot of the close relationship between body length and generation time across organisms ranging in length from less then one-thousandth of a centimetre to almost one hundred metres and with generation times that vary from minutes to decades (Fig. 1.1). Could the fact that the relationship is approximately linear when both axes are logarithmically scaled be related to the fact that growth is essentially a logarithmic process? If so, why should organisms of similar size sometimes have quite different generation times (compare the mouse with the horseshoe crab)? Can we think of differences among similar-sized species which might help explain this variation?

Maximum population densities of different species are closely correlated with body mass (Damuth 1987; Fig. 1.2). Across a range of taxa varying in size from viruses and bacteria, to sequoia trees and whales, larger-bodied

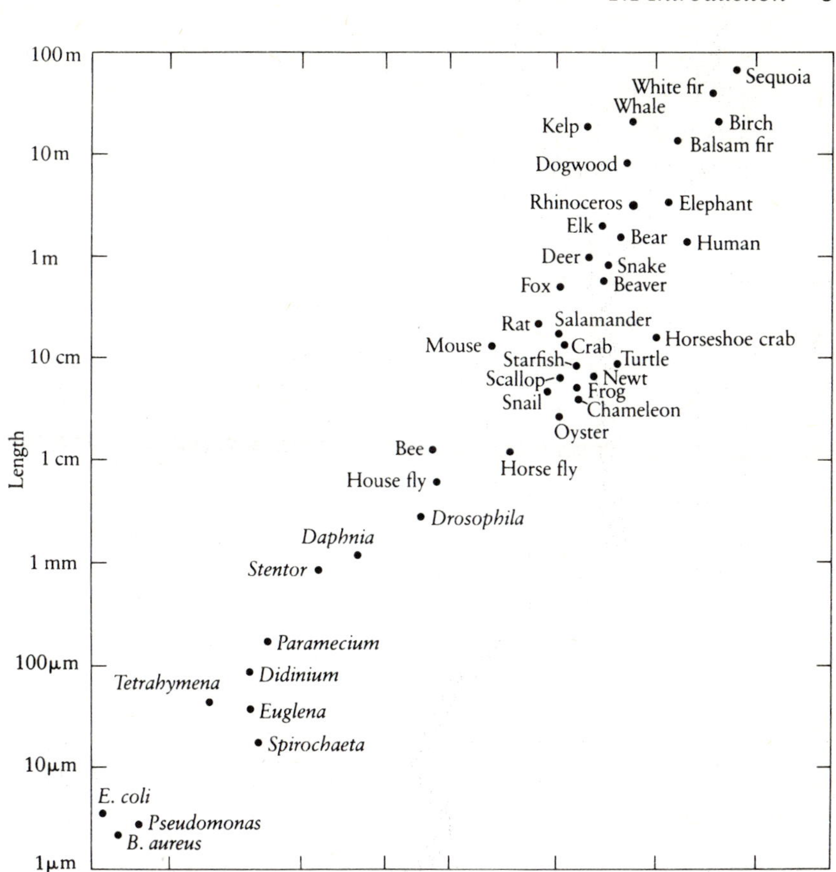

Fig.1.1. The relationship between generation time and body length across species ranging in size from bacteria to sequoia trees and whales. (From McMahon and Bonner 1983, after Bonner 1965).

species live at lower population densities. Perhaps the fact that heavier organisms need more resources to maintain themselves, grow, and reproduce, means that a given area of habitat can sustain fewer of them? However, when different species' energy needs are considered together with the data in Fig. 1.2, a surprising finding emerges. In a given area, the population of each species which exists at its maximum population density uses approximately the same amount of energy (Damuth 1981, 1987). The ecological process that results in populations of different species obtaining approximately equal amounts of energy remains unknown.

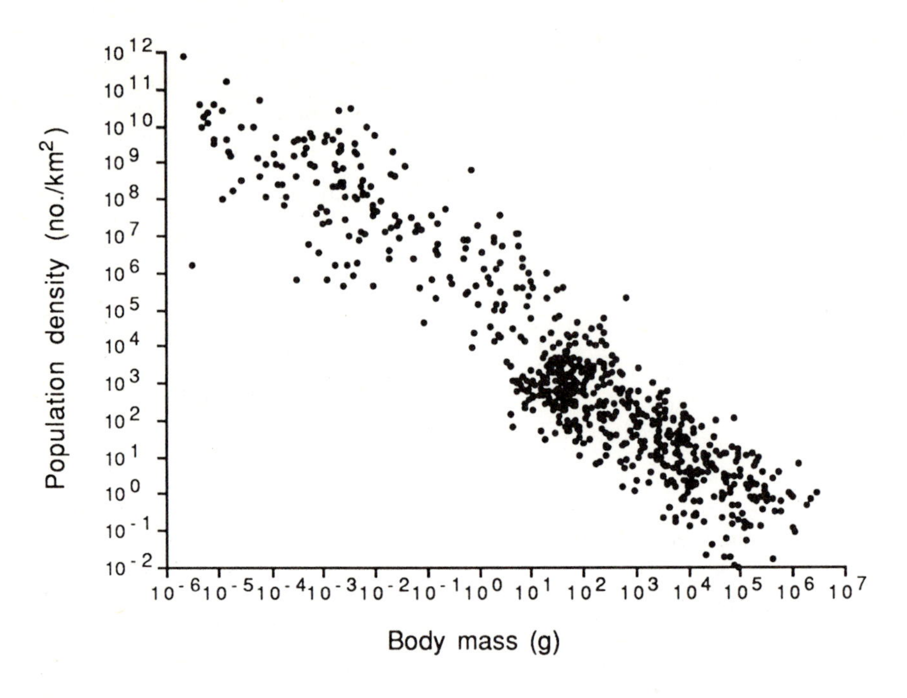

Fig. 1.2. The relationship between population density and body mass across species ranging in size from viruses to sequoia trees and whales. (Data from Damuth 1987).

In addition to the sort of regularities shown in Bonner's and Damuth's comparisons, further patterns are often revealed on closer inspection of the data. For example, testes weight increases with body weight across primate species. However, those adult male primates that belong to species with large testes for their body weight are the ones living in social groups containing several reproductively active males (Fig. 1.3). Why should this be? The answer lies with mating patterns adopted by females, a point we shall return to later in this chapter.

Finally, comparative studies can demonstrate a lack of variation in some characters, despite wide variation in other characters that we might have expected to be correlated with them (see Stearns 1984). For example, for bird and mammal species living in a wide range of social group sizes and with mating patterns that seem to span the range of what is possible (monogamy, polygyny, polyandry, polygynandry, and even promiscuity), the frequency of inbreeding between parent and offspring or between sibs is generally of the order of 1 or 2 per cent of all matings (Fig. 1.4). The exceptions tend to be from populations where mate choice is severely

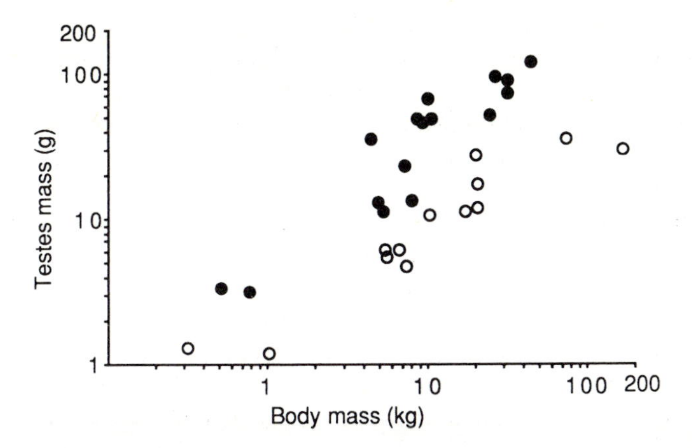

Fig. 1.3. The relationship between testes mass of adult males and body mass across primate species. Males belonging to species in which females are likely to copulate with more than a single partner per oestrus (●) have larger testes for their body weight than those where females invariably mate with only a single male (○). (Data from Harcourt *et al* 1981 and Harvey and Harcourt 1984, with additional material from Terborgh and Goldizen 1985.)

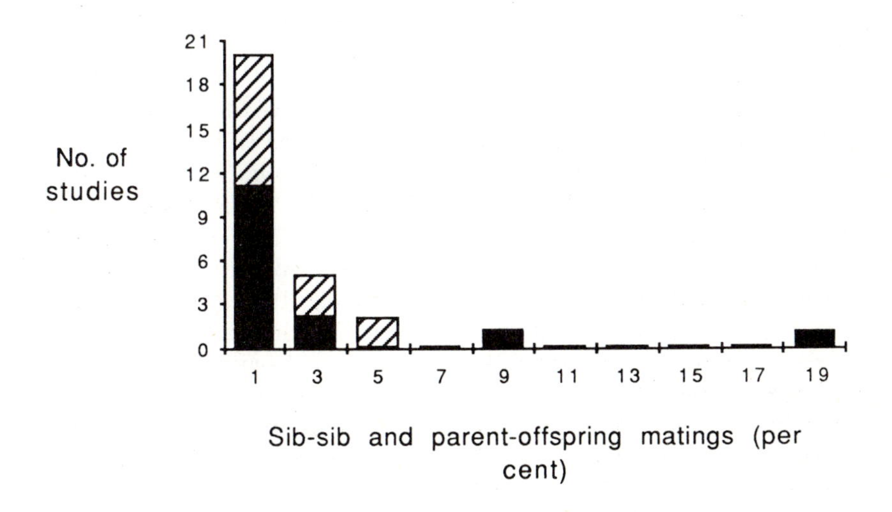

Fig. 1.4. Percentage of matings that are incestuous in populations from different species of birds (solid bars) and mammals (hatched bars). One main feature in the graph is the low frequency of such matings in the vast majority of species. The two outliers are small populations in which mate choice is extremely limited. (After Harvey and Ralls (1986) and Ralls, *et al* 1986.)

limited (e.g. a breeding population of five swans). There are costs associated with inbreeding because the resultant offspring tend to be homozygous, and the products of rare deleterious recessive genes are thereby expressed. Other things being equal, potential parents would do well to avoid incest. Other things are not always equal, however, and incestuous matings occur. One question posed by Fig. 1.4 is why should the level of inbreeding in birds and mammals generally be of the order of about 1 per cent rather than 0.01 or 10 per cent?

Although we shall often focus on identifying adaptive trends, the absence of correlations between character states and environmental differences may suggest non-adaptive interpretations. Furthermore, we shall describe examples to demonstrate that the repeated evolution of the same character states in similar environments need not necessarily have an adaptationist interpretation. We must relentlessly seek alternative explanations, and we must entertain them seriously, whether they invoke physical laws or genetic constraints. Similarly, we must always be ready to abandon favoured taxonomies, to re-classify phenotypes and environments, and to reconsider tried and tested theories. We have used examples liberally in many parts of this book, hoping that some at least will strike familiar chords with the reader. These are exciting times for comparative biology because most of the issues are clear, much of the groundwork has been laid, and the data are accumulating at least as rapidly as the scientific community can make sense of them.

This book does not attempt an historical survey of the comparative method's many accomplishments. That would have meant reviewing almost all of Darwin's many writings including *On the origin of species by means of natural selection* (Darwin 1859) and *The descent of man and selection in relation to sex* (Darwin 1871); critically summarizing several other landmark texts including D'Arcy Thompson's (1917) *On growth and form* and Julian Huxley's *Problems of relative growth* (Huxley 1932); evaluating the field of animal mechanics (Alexander 1968, 1982; McMahon 1983); and explaining how an amazing diversity of behaviour, morphology, and mating patterns has been related to simple ecological differences in several animal taxa, from weaver birds (Crook 1964), through other birds (Crook 1965; Lack 1968), to ungulates (Jarman 1974), and primates (Crook and Gartlan 1966; Clutton-Brock and Harvey 1977). Even then we should merely have scratched the surface[1].

Instead, as we mentioned above, we have in mind the new rigour that has pervaded the field of comparative biology for a little more than a decade. It matters not at all whether you work with genetic elements, with

---

[1] A partial but informative and entertaining history of comparative studies is given in the first chapter of Ridley (1983*a*).

viruses, bacteria, fungi, animals, or plants. The same principles apply if your subject is molecular evolution, the diversity of genetic systems, comparative morphology, physiology, ecology, or behaviour. If you are to interpret organic diversity correctly using comparisons, you need to think carefully about the methods you might use. This book is about those methods and the assumptions that they make.

1.2 The organization of this book

The book consists of seven chapters. This first chapter identifies the sorts of problems that the comparative approach can answer, and shows how it complements other approaches to problem-solving in evolution, such as optimality theory, population genetic models, and experimentation.

The second chapter identifies the biological causes of the most important problem facing comparative biologists, that of similarity among closely related species for almost any character that we look at. Closely related species often inherit traits from common ancestors. Treating species as independent points in statistical analyses may, therefore, greatly overestimate the true number of degrees of freedom. However, if we do not know why closely related species tend to be similar, our arguments will be based on statistical models that have no firm biological foundations—an unsatisfactory state of affairs.

Because the key to comparative analyses depends on understanding the phylogenetic relationships among the sample of species being considered, the third chapter discusses methods for reconstructing phylogenetic trees and ancestral character states. The importance of defining ancestral character states is that they allow us to estimate the amount of evolutionary change in each branch of a phylogenetic tree. This chapter will illustrate the importance of the assumptions we make about the way evolution proceeds: different models of evolution can produce quite different phylogenetic trees and ancestral character states.

On the assumption that phylogenetic trees and ancestral character states are known, the fourth chapter sets out to develop statistical tests that will determine whether different characters that exist in discrete states show evidence for correlated evolution. For example, caterpillars have sometimes evolved from being cryptic to being warningly coloured and, on occasion, they have also become distasteful to predators. Have the two traits, warning coloration and distastefulness, evolved independently of each other? After reviewing alternative methods, we develop a more general model for the evolution of characters that change state from time to time. That model, which assumes that characters evolve independently, can be used as a null hypothesis against which to test real data which, we might suspect, can demonstrate that correlated evolution has occurred. To pursue the noxious caterpillars for a moment, we might demonstrate that the phylogenetic ancestry of warning coloration and distastefulness does not accord with our model of the two characters evolving independently. We then return to the alternative methods that have been developed to detect correlated evolution between discrete character states. We show how they can be viewed as special cases of our more general model which make particular assumptions about branch lengths in phylogenetic trees and about similarity in rates of character change.

The fifth chapter turns to comparative analyses of continuously varying characters. On the basis of such characters, every species differs, however minimally, from every other species. As a consequence, there must have been change in the character states along all branches of the phylogenetic tree that relates the species in a sample. Two problems are posed by such characters. We must first define independent comparisons for statistical testing, and then we must devise appropriate tests that can detect correlated evolution. Over the last few years, many ways of seeking independence have been suggested and a number of tests have been devised. We assess the different tests in the light of the evolutionary and statistical models they are based on and we conclude that one particular approach, that of independent comparisons either between species and higher nodes or along the branches of the phylogeny, is the most satisfactory. Again, we stress the relationship between statistical models and the process of evolution. Each statistical model makes assumptions about the way evolution has occurred, and the choice of a statistical method is really the choice of a model of evolution.

The sixth chapter deals not with *whether* characters have evolved together, but with the *way* in which they show correlated evolution. The form of the relationship between two continuously varying characters can suggest to us reasons for the relationship. Allometry provides good examples because many characters vary with body size in ways that are dictated by physical processes. We shall describe a general statistical model that can be used to identify the forms of allometric relationships, and show how the more usual procedures for line fitting are special cases of the general model, each making different assumptions about sources of variation in the sample of organisms used. Those assumptions may or may not be valid in any particular case. We also show how independent evolutionary occurrences of functional relationships, such as allometric relationships, can be identified and used to estimate the general relationship. Finally, we discuss statistical models for identifying allometric relationships when the variable to be predicted is represented as a function of two or more predictor variables.

The seventh and concluding chapter argues that comparative analyses will be used more widely in molecular biology on the one hand and ecology on the other, but that many future developments of comparative methodology await more accurate understanding of evolutionary processes.

Comparisons are not made simply to help understand adaptation. In the next section of this chapter (Section 1.3) we discuss how the comparative method works at the interface between the two classic traditions in comparative biology—reconstructing phylogenetic trees and studying adaptation. We then examine more carefully the different types of question that comparative biologists might attempt to answer, and the types of comparison that can be used to answer them (Section 1.4). In particular, we discuss how adaptation can be inferred from comparative studies, and we distinguish between the two main types of comparative study which differ in the comparisons made. Directional studies compare ancestors with descendants, whereas non-directional studies compare character states among daughter taxa without considering explicitly the character state of their most recent common ancestor.

Having laid the groundwork for comparative studies, we then show how they can complement other approaches to problem-solving in evolutionary biology (Section 1.5). Comparative studies can be based on both the kinematic models of population geneticists and the optimality approaches favoured by behavioural ecologists. We use the two case studies of sex ratio and home range evolution to illustrate the complementary role of the comparative, experimental, and observational methods for tackling evolutionary problems (Section 1.6).

We conclude this introductory chapter with a short account of the atomization of characters and environments (Section 1.7): organisms can be butchered in many ways, but some cuts are more natural than others.

1.3 Two traditions in comparative biology: descent and guilds

There are two traditions of comparative biology (see Ridley 1983*a*), which might be called the descent and guild schools. For the most part, taxonomists belong to the descent school, whereas ecologists belong to the guild school. Taxonomists search for natural ways of classifying organisms and phylogenetic relationship forms the obvious unifying principle: organisms are classified on the basis of common ancestry[2]. Ecologists, on the other hand, recognize guilds as groups of animals that share a common way of life (Root 1967). Members of a guild may be close phylogenetic

---

[2] There are schools of taxonomy which are not based on classification by common ancestry. We shall describe the most influential of them in Chapter 3.

relatives, but frequently they are not. Convergent or parallel evolutionary change (Fig. 1.5) can lead to phenotypic similarities among the members of a guild that are not close relatives.

These two types of comparative biology need to be brought together. We make comparisons because we want to understand organic diversity, and that usually means unravelling the reasons for evolutionary change and stasis. Any difference among organisms, whether the trait is labile over evolutionary time or not, may eventually turn out to have an adaptive basis. Accordingly, it is important to ensure that procedures to incorporate information on phylogenetic relatedness into comparative tests do not rule

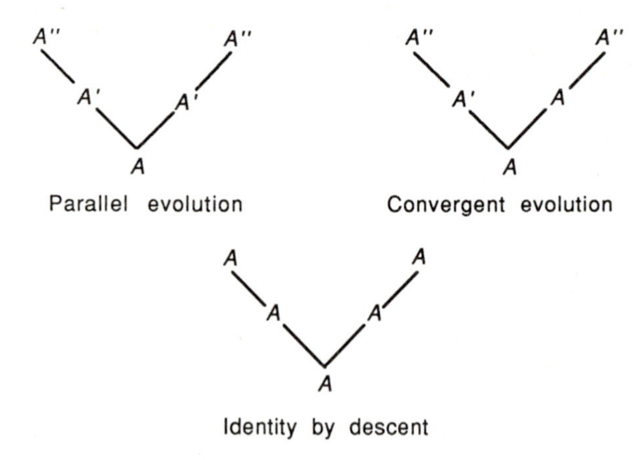

Fig. 1.5. Three phylogenetic trees showing the evolution of a single character which may occur in any one of three states: A, A', and A''. If each character state is unique to a particular environment (say A with E, A' with E′, and A'' with E″) then both parallel evolution (the same phenotypic change occurring in separate lineages: A′ to A″ in the figure) and convergent evolution (the same phenotype arising from phenotypically different ancestors: A and A′ both changing to A″ in the figure) may suggest evidence for evolution of adaptation by natural selection. If the environment has remained the same, as in the identity by descent case, adaptation still cannot be ruled out (phenotype A may be adapted to environment E). Separate evolutionary origins of the same character states, under either convergent or parallel evolution, define them as analogous character states. The single origin of the character state under identity by descent labels it homologous in the two taxa. We shall discuss in Chapter 3 how ancestral character states are reconstructed, but this figure presages one of the problems highlighted in that chapter. If the only information available is the character states of the pairs of extant species, and the most parsimonious phylogenetic trees are sought, then each tree would demonstrate identity by descent: the common ancestor of two species with identical character states would also be deemed to have had that phenotype. The frequency of convergent and parallel evolution would be under-estimated, while the frequency of identity by descent would over-estimated.

out the possibility that identity by descent from a common ancestor is maintained by selection. Evolutionary biologists have demonstrated many instances in which identity by descent is, indeed, maintained by selection (see Cain 1964; Endler 1986). In the next chapter we shall provide an explicitly adaptive scenario for the maintenance of identity by descent (Section 2.3.1).

1.4 The comparative approach for studying diversity

Comparative studies identify evolutionary trends by comparing the values of some variable or variables across a range of taxa. The variables may include descriptions of the environments inhabited by the organisms as well as phenotypic characters. Huey (1987, p. 76) describes these procedures as 'documenting the extent and pattern of organic diversity'.

We have been careful to avoid saying that comparative studies are only concerned with the study of adaptation. As we shall see throughout this book, comparisons do often help us to understand the adaptive significance of phenotypic variation. Stripped to the bone, however, the evidence for adaptive evolution revealed by comparative studies is correlated evolution among characters or between characters and environments. Nevertheless, such evidence can be convincing, and the success of one comparative test can lead to others that produce a better understanding of the reasons for organic diversity. Our goal is to identify the variable or variables responsible for variation in some other variable. For example, taxonomic variation in testes size among mammals may reflect the influence of two such very different variables as the production of testosterone and the production of spermatozoa. We can use variation in testes size as an example of how comparative studies often proceed.

Primates and other mammals with larger body weights generally have larger testes. This pattern might be expected for at least two reasons. First, larger bodied species need larger endocrine glands to maintain threshold levels of testosterone in the blood. Second, if larger bodied species are to fertilize an egg successfully after mating, they need the capacity to produce more spermatozoa than smaller species if they are to counter the dilution effects of a larger female reproductive tract. However, there are interesting comparisons which provide exceptions to the positive relationship between body weight and testes size. For example, the gorilla *Gorilla gorilla* is four times the body weight of a chimpanzee *Pan troglodytes*, but has testes which are one quarter the mass of a chimpanzee's. In a series of articles, Short (1977, 1979, 1981) developed the idea that the difference in relative testes size between the chimpanzee and the gorilla was a consequence of sperm competition. Female chimps regularly mate with several males during a single oestrus, and males can increase their chances of paternity by producing more sperm (like having more tickets in a lottery), hence the larger testes. Female gorillas mate with only one male, and sperm competition is therefore not important.

The sperm competition idea makes a testable prediction for other primates. Harcourt *et al.* (1981) predicted that those primates with relatively large testes would be the ones living in multi-male groups where females had the opportunity of regularly mating with more than a single male during a given oestrus. The prediction proved to be correct (Fig. 1.3). But the comparative tests did not stop there. Subsequent work on primates shows that those sexually selected traits having to do with competition for access to mates vary in predictable ways with testes size (Harvey and Harcourt 1984), and that rates of sperm production are higher in species with relatively large testes (Møller 1988*a*). Furthermore, analyses both within other orders of mammals (Møller 1989) and among birds (Møller 1988*b*) reveal similar relationships between testes size, sperm production and mating patterns (for a review, see Harvey and May 1989).

In the testes example, the pattern of variation suggested to Short a possible adaptive explanation. The first test of the explanation involved controlling for extraneous influences (in this case body size) while varying the presumed causal influence (the amount of sperm competition). An alternative to using statistical methods (such as regression analysis or partial correlation) to control for the effects of body size would have been to compare many species of the same size (Smith 1980) but, because of the relatively few species involved, that did not prove practical for primate testes. Subsequent tests involved other taxa and finer-grained comparisons which focused on other correlates of testes size, such as sperm production. The example shows how, once patterns and probable causal variables have been identified, comparative studies can begin, cautiously, to make inferences about adaptation. However, because comparative studies seldom have access to the actual selective forces, inferences about adaptation are partly a matter of good comparative methodology. The demonstration that similar relationships evolved in different families of primates, different orders of mammals, and in birds, lends additional credence to Short's explanation because it is unlikely that the same alternative explanation could be responsible for an association between mating system and relative testes size in all those taxa.

Other comparative studies can provide strong evidence against particular adaptive explanations for species differences. One example concerns life history variation among vertebrates. Some vertebrates seem to live faster lives than do others: even after body mass has been factored out by partial correlation, short gestation lengths are associated with early ages at maturity and short reproductive lifespans. It has been suggested that those species which live relatively fast lives are able to do that because they have higher metabolic rates (e.g. Brody 1945; McNab 1980, 1986*a*, *b*). When this idea was tested in both birds and mammals, there were no statistically significant correlations between speed of life and rate of metabolic turnover after controlling for differences in body mass (Harvey *et al.* 1990; Trevelyan *et al.* 1990).

In the next section we develop further some of these ideas about the methods and products of comparative studies. Immediately below we discuss what can and cannot be concluded from simple comparative relationships. We have in mind here specifically the inference of adaptation from comparative data. We argue that the inference of adaptation is often an explicitly historical one, that depends on being able to say something about the transition from ancestral to derived conditions. Nevertheless, as the testes example showed, it will often be possible to argue that differences are adaptive without knowledge of ancestral conditions.

The subsequent section (Section 1.4.2) will examine further the difference between comparisons of ancestral and derived conditions (directional comparisons) and comparisons among taxa that do not specify ancestral and derived conditions (non-directional comparisons). Methods appropriate for examining these two different approaches will be discussed in Chapters 4 and 5.

1.4.1 Inferring adaptation from comparative studies

If comparative studies reveal only correlated evolution, how do we go about inferring adaptation from comparative relationships? We shall first explain what we mean by an adaptation, and then examine how the concept can be applied to comparative studies.

What is an adaptation? The answer to this seemingly straightforward question has been and will continue to be debated in the literature (e.g. Mayr 1982). Adaptation is an inherently comparative idea (see also Hinde 1975; Clutton-Brock and Harvey 1979). When we ask how white rabbits are adapted to snowy environments, our answer will inevitably make an implicit or explicit comparison with rabbits that are not white. For example, we might answer that: 'White rabbits are adapted to snowy environments because they are camouflaged against a background of snow, and are thereby protected against predators', by which we mean that, for rabbits living in a snowy environment, white provides better camouflage than other colours. Similarly, when we ask why leaves are an adaptation for flowering plants, we have in mind a comparison with plants that do not have leaves. However, we shall adopt a more restricted meaning of the term: for a character to be regarded as an adaptation, it must be a derived character that evolved in response to a specific selective agent. The rabbit's white coat would be an adaptation for camouflage if it evolved from a brown ancestral condition in response to selection for camouflage in snowy environments. If, however, it evolved from brown because of its improved thermal properties, it would not be an adaptation for camouflage but, perhaps, an exaptation for camouflage (Gould and Vrba 1982). How, does this notion of adaptation fit in the Darwinian scheme of things?

Although evolution by natural selection can produce adaptations, the concept of adaptation is not necessarily inherent in a description of natural selection. For example, Lewontin has often argued that evolution by natural selection occurs when: (1) there is phenotypic variation; (2) that variation is heritable; and (3) some variants leave more reproductive offspring than others (e.g. Lewontin 1978).[3] To introduce the idea of adaptation, several authors (e.g. Williams 1966; Brandon 1978; Dunbar 1982; Krimbas 1984; Coddington 1988) would modify this scenario to the effect that: (3) some variants function better than others and are thereby better adapted; and (4) the better adapted variants leave more offspring. To avoid circularity, we need to explain *why* some variants function better than others, and that accords with our understanding of adaptation. For example, white rabbits may leave more offspring than brown rabbits in snowy environments, and the reason *why* white rabbits leave more offspring is that they live longer than brown rabbits because they are less likely to be detected by predators. The process is specific to a particular environment, so that different variants may be the better adapted in different environments; brown rabbits would be better camouflaged in snow-free woodlands. Evolution by natural selection is also specific to a particular ancestral state: white rabbits must have evolved from something. The natural comparison, then, is between ancestral and derived traits.

According to the above perspective, adaptations are produced by natural selection. Coddington (1988, p. 5) comes near to our meaning of an adaptation, which he defines as 'apomorphic [evolutionarily derived] function due to natural selection'. If we were interested in finding out whether aposematic coloration among insects evolved to advertise distastefulness (Harvey *et al.* 1982) we should look for several different instances of the origin of distasteful aposematic insects from cryptic palatable ancestors. If, whenever distastefulness evolved in lineages of cryptic palatable insects, aposematic coloration soon followed, we should see that as useful support for our adaptive generalization.

The methodology that Coddington proposes for studying adaptation comparatively is summarized in Fig. 1.6. Character state 1 is shown here to have evolved in the branch linking the nodes that lead to species *B*

---

[3] Heritability of traits will usually differ among environments (Falconer 1981; Lewontin 1974, 1982), and no distinction need necessarily be made between genetic and cultural transmission (Boyd and Richerson 1986).

and C. Character state 1 is then lost in the branch leading to species E. The adaptive hypothesis is that 1 evolved from 0 due to natural selection, and to perform derived function 1, denoted F_1. Implicit in the hypothesis is the belief that 1 evolved because it conferred an advantage to its possessors for performing F_1, compared to individuals with 0. Adaptation is here defined explicitly with respect to a primitive feature. This rules out the question 'What is 0 in species A and B an adaptation for?', at least with respect to the phylogeny given in Fig. 1.6. Even though 0 may be derived in these species with respect to some more primitive state, such a state is not represented in the phylogeny of Fig. 1.6.

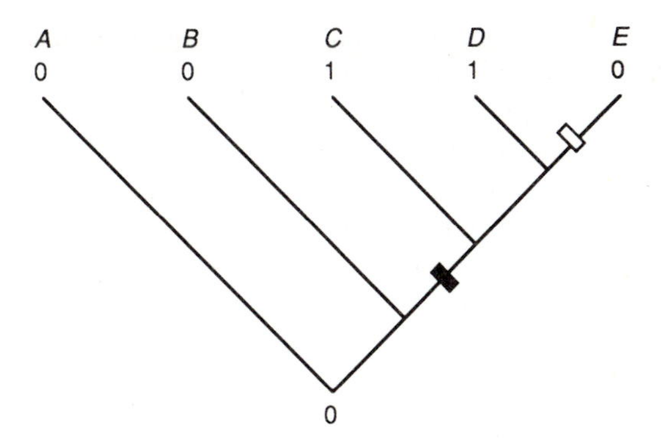

Fig. 1.6. Five species, labelled A to E are scored for a character which is either in state 0 or state 1. The dark cross bar indicates the acquisition of 1 from 0, and the open cross bar is the reversion of state 1 to 0. The hypothesis is that between the nodes leading to B and C, character 1 evolves from character 0 in response to natural selection. (After Coddington 1988).

The secondary loss of 1 in species E produces a different origin of 0 which may be an adaptation with respect to 1. However, the evolution of 0 in that species follows rather than precedes the evolution of 1, and thus 0 in species E is not ancestral to 1. Thus, Coddington's methodology defines which taxa are to be used as 'control' groups for testing the adaptive hypotheses. Species E in Fig. 1.6, despite lacking 1, is not appropriate for testing whether 1 is an adaptation with respect to 0.

Coddington's strict cladistic approach to studying adaptation may seem restrictive to those used to studying adaptation by comparing the character states of groups of species or higher taxa that inhabit different environments or which differ in other ways thought to be responsible for variation in the trait of interest. Put another way, what is wrong with studying adaptation by assessing the fit or current utility of characters to their environments? The answer to this question hinges on a subtle but important distinction. If we believe that character variation is largely due to the effects of natural selection, then it is reasonable to suppose that variation in complex characters across current environments does represent *adaptive* variation in the sense that the character states associated with particular environments serve a function. We do not usually have an alternative theory to explain recurrent similar fits between a trait and the environment. However, analyses of current utility do not necessarily inform us of the origin of trait variation. That is, the associations of traits with environments does not necessarily imply that those traits are *adaptations* to perform particular functions in those environments, where by an 'adaptation to perform function X' we mean a character that arose by natural selection to perform function X.

Some examples may help make these points. Suppose we were interested in whether long necks in mammals had evolved to help them browse from the leaves of trees because grass was in short supply. If we sought a correlation between leaf eating and long necks across mammals, it is perfectly possible that no significant relationship would be found. This is largely because the lack of long necks need not indicate the lack of selection pressure to feed on leaves: many mammals climb up trees to feed on leaves. Use of the climbing species as controls to test for long necks as an adaptation for leaf eating makes the same error as using species *E* in Fig. 1.6 as a control for species *C* and *D*. Species *E* may lack the character for reasons that are different from those in the ancestral condition of species *A* and *B*.

The proper comparative test in this instance would be to compare long-necked mammals with their short-necked ancestral forms. If it were also possible to argue that the short-necked ancestors of long-necked forms, such as the giraffe, lived in treeless environments whereas the long-necked forms always fed from trees, then long necks in the descendants would constitute evidence for adaptation.

Surely, though, if we actually find a relationship between character and environmental variation among contemporary forms we can give it an adaptive explanation? Here the argument has both semantic and plausibility components. Suppose, for example, that we find longer canine teeth in the adult males of primates species with intense male–male competition, and that the males of species lacking such competition also lack long canines. It seems reasonable to argue that long canines are an adaptation to male–male competition, which they probably are (Harvey *et al.* 1978). The criticism of such a conclusion, however, is that nothing in the 'current utility' methodology guarantees that long canines are the derived form. It seems very likely that they are, but if they are not it makes no sense to talk of long canines as an adaptation, with respect to short canines, for use in male–male competition. This may seem a mere semantic point but consider our position if males with long canines predated the origin of intense male–male competition among primates, we would have to give a somewhat different explanation for their present day association with mating systems.

This latter point brings up the plausibility argument. It is implausible that a costly character like long canine teeth could be *maintained* without having some function. The argument goes that if long canines had no function they would be lost by natural selection over evolutionary time. So, if we find them repeatedly in species with male-male competition, and if it can be shown that similar levels of male-male competition are not found in primates in which the males lack long canine teeth, some selective explanation is called for. The argument is slightly different from that which links the *origin* of a character to a particular selective force. Here, even if the origin of a character cannot be attributed to a particular selective force, it is suggested that it is maintained by it. Regardless of their origin, the adult males' long canines have a function. In this sense, *variation* in canine length among species is adaptive even if the long canines are not adaptations in Coddington's sense of the term. Gould and Vrba (1982) would label long canine teeth in this context an exaptation. This is a useful term in so far as it calls our attention to the fact that the particular solution to an environmental problem may depend on what selection has to work with. The following (unlikely) scenario illustrates the point. Long canines evolved in the adult males of ancestral primates as a result of selection to defend their mates and young against predators (DeVore and Hall 1965). The primates subsequently have evolved societies with intense male-male competition and the long canines were used in combat between males. The long canines are exaptations. (We should point out, however, that if canines became even longer in response to selection for fighting than they were to ward off predators, the difference is an adaptation.) The original predators now went extinct and, as the primates radiated, those lineages in which male-male competition became less intense evolved shorter canines because males with long canines were less fit. The short canines are the adaptation.

Coddington (1988) provides a nice example that reinforces the points we are making about the current utility of a character. An accepted dogma in arachnology was that the orb web evolved from a primitive cob web as an adaptation for catching flies more effectively. However, careful cladistic analysis indicates that orb webs were ancestral to cob webs (Coddington 1986*a,b*), so the proper question is what were cob webs an adaptation for? If an analysis of current utility shows that, indeed, orb webs are more efficient than cob webs at catching flies, how should the variation be interpreted? Originally, we might have supposed that the correct mutants to allow the evolution of orb webs from cob webs had not occurred but that if they had then orb webs would have evolved in cob web lineages. The cladistic analysis argues against that interpretation.

We have already mentioned the problem that environmental correlates of phenotypic variation may result from the confounding effects of some third variable which does not have a direct effect on (nor is directly affected by) the character state or trait in which we are interested. But, even when there is a causal relationship between the environmental variable and character state, does the relationship have to be adaptive? The answer to this question is no.

One reason why adaptive traits have apparently failed to evolve in some situations is because of genetic constraints. For example, toxic compounds are frequently sequestered by caterpillars from their food plants, thus rendering those species feeding on poisonous food plants unpalatable. Caterpillars do not synthesize such toxins, which means that caterpillar species living on palatable food plants have not evolved to become distasteful (see Rothschild 1972). If ancestral caterpillars lived on toxic food plants and were unpalatable, but we could identify several instances of the evolution of palatability associated with a switch to host plants that are not toxic, the evolutionary association might well not be adaptive. If caterpillars living on palatable food plants could have evolved to become toxic to their predators, they probably would have done.

There have been many useful criticisms of the so-called 'adaptationist program' (e.g. Lewontin 1978, 1979; Gould and Lewontin 1979). The comparative method might seem to fit nicely into the mould defined by the adaptationist program, which was caricatured by Lewontin (1979): (1) find phenotypic variation; (2) ascribe genetic causation to that difference; and (3) produce an adaptive explanation for the difference by, for example, 'imaginative post-hoc reconstruction'. That procedure is, indeed, adaptive story telling, and the trap is all too easy to fall in to. Our procedure runs differently: (1) find phenotypic variation among taxa; (2) produce one or more adaptive explanations for that variation which may include assumptions about heritability and ancestry; and (3) test the explanation(s) by predicting particular environmental or constitutional correlates of the variation and by comparing ancestral and derived character states wherever possible.

To summarize this section. Comparisons between contemporary forms, unsupported by attempts to reconstruct ancestral character states, often reveal correlations between character states or between character states and environments that can readily be given an adaptive interpretation. Not all such differences are, however, adaptive and we must be careful to examine alternative explanations. The proper inference of an adaptation *per se* depends critically on understanding which features are primitive, and which derived. Many characters have different functions from those for which they evolved. As a consequence, we should not too readily label characters as adaptations to their current function, even though they may confer current selective advantage. Nevertheless, changes in such characters from their original state can be thought of as adaptive changes.

1.4.2 Directional and non-directional comparisons

Two different but complementary trends have recently emerged in comparative studies (Huey 1987; Pagel and Harvey 1988*a*). What we shall call 'directional' studies make use of ancestral character states to infer the direction and rates of evolutionary change between ancestors and descendants. Not surprisingly, in the absence of a good fossil record, this branch of comparative studies draws on developments in the reconstruction of phylogenies and ancestral character states. In contrast, 'non-directional' studies analyse evolutionary trends across either contemporary species, or across higher nodes which are usually at a similar taxonomic or phylogenetic level. In the past, non-directional studies did not make much use of phylogenetic information. However, the picture is changing rapidly as new developments in the analysis of contemporary forms which do rely on phylogenetic information have become available.

Stated more simply, the two different approaches to comparative analysis can be thought of as looking down lineages over time versus looking across different lineages (Fig. 1.7). The distinction between the directional and non-directional approaches is more than just conceptual. Each approach has given rise to one or more comparative techniques, some appropriate to discrete variables, some appropriate to continuous variables. These techniques are the topics of Chapters 4 and 5.

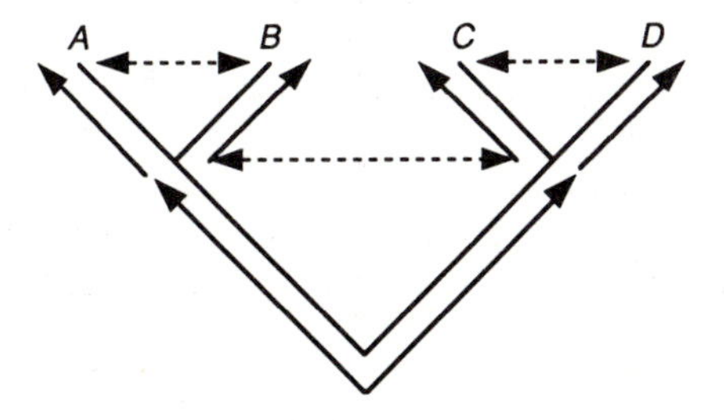

Fig. 1.7. The difference between directional and non-directional comparisons. The phylogenetic tree shows the relationships among the four extant species *A*, *B*, *C* and *D*. Species A and B are in one genus, with species *C* and *D* in another. The solid arrowed lines represent directional comparisons between ancestors and descendants, while the dotted arrowed lines represent non-directional comparisons between taxa that each have part of their ancestry independent of the other.

Huey (1987) and Huey and Bennett (1987) report an example of a directional analysis of changes in preferred body temperature and optimal running temperatures in lizards. These authors were interested in the idea that changes in optimal running temperatures over evolutionary time have kept pace with changes in preferred body temperatures. If they have, then the differences between ancestors and descendants, measured over many different branches of a phylogeny, should covary with a slope of 1.0. In fact, they found that this slope was significantly less than 1.0, indicating that changes in optimal running temperatures have not kept up with changes in preferred body temperatures in these lizards. Because directional analyses rely on explicit reconstruction of ancestral states, these authors were also able to describe the direction of evolutionary change. Over time lizards have evolved to run at lower preferred temperatures. We shall describe in more detail in Chapter 5 the method that Huey and Bennett (1987) used.

Directional tests are also available for discrete characters. Some studies have simply asked whether particular evolutionary sequences among states of a single character are more likely than others. For example, Gittleman (1981) tested the suggestion that the only types of transition likely between patterns of parental care in bony fish would be between: (i) no care and paternal care; (ii) no care and maternal care; (iii) paternal care and biparental care; and (iv) maternal care and biparental care (Fig. 1.8). Of 21 transitions, all were of the predicted type and none were between: (v) biparental care and no care; or (vi) paternal care and maternal care. Similarly, Carpenter (1989) tested West Eberhard's (1978) rather complex model for the evolution of patterns of social behaviour in the vespid wasps. Some transitions did not accord with prediction, although others did.

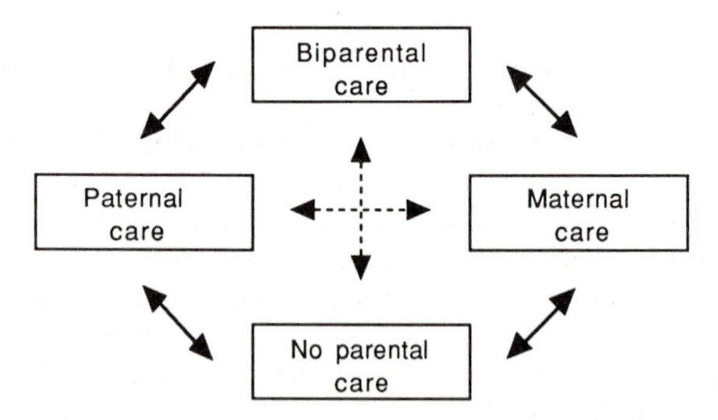

Fig. 1.8. Directional tests of parental care in fishes. Four states are possible. Solid arrowed lines are hypothesized transitional routes, whereas dotted lines do not accord with the hypothesis. All 21 transitions detected by Gittleman (1981) were in accord with the hypothesis.

Other directional tests are available to examine the correlated evolution of two or more characters. Donoghue (1989), using a method developed by Maddison (1990), tested whether the evolution of dioecy in plants was more likely to occur in plants with fleshy or with dry propagules. There was only partial support for the hypothesis. Donoghue's results, and Madison's method are described in Chapter 4.

The more traditional approach to comparative studies involves non-directional analysis of contemporary forms. As will be seen in Chapters 4 and 5, a variety of methods have been developed for this kind of analysis. Many of the now classic studies of allometric scaling are examples of this approach. More recently, several new techniques have been developed that measure pairwise differences between sister taxa at all levels of a phylogeny. These techniques, described primarily in Chapter 5, offer much hope for extracting from a comparative data set a group of independent comparisons with desirable statistical properties, each of which bears on the comparative idea being tested.

Superficially, non-directional analyses would seem not to make as much use of phylogenetic information as do directional analyses. This has been true in the past. However, non-directional analyses now rely to the same extent as directional methods on patterns of phylogenetic branching, and on the reconstruction of ancestral characters. Where directional analyses make use of this information to examine directions and rates of evolutionary change, non-directional analyses examine the nature of covariation among different phylogenetically defined groups. Sometimes such relationships vary depending upon the taxonomic or phylogenetic level. Huey (1987), for example, summarizes several studies on the relationship between 'performance breadth' (the range of body temperatures at which lizards can exceed some level of performance), and the range of body temperatures at which they are commonly found in nature. Across genera these two traits are negatively correlated. However, among more closely related taxa, the expected positive relationship holds.

The choice between the two sorts of methods, then, is a choice between the kinds of ideas one wants to test. Directional tests offer much promise as techniques for phylogeny and ancestral character state reconstruction improve. Furthermore, they test directly the transition from primitive to derived forms, and so are more immediately responsive to the criteria outlined in the previous section for studying adaptation. Directional tests may prove to be very useful for detecting instances of parallel and convergent evolutionary change (Fig. 1.5). In some instances parallel change may be masked if non-directional analyses are used (Fig. 1.9). Non-directional analyses of contemporary forms are particularly useful for detecting and describing the nature of the current utility or fit between characters and environments across taxa. Coupled with careful biological

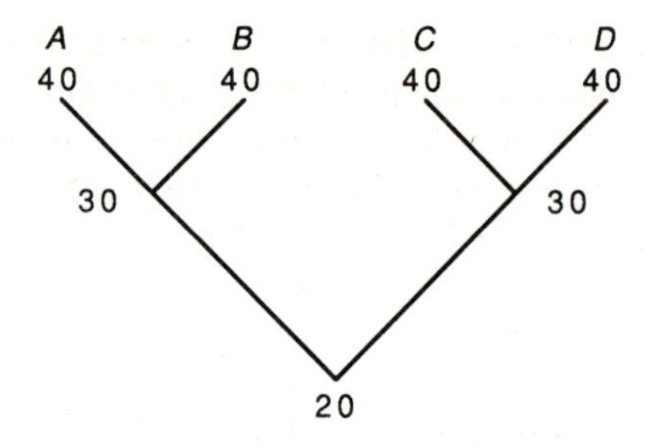

Fig. 1.9. The same phylogeny as Fig. 1.7 but with the state of a particular character, such as body weight, given for each species and node. Directional analysis would reveal parallel evolutionary change, or Cope's Law (Cope 1885; Newell 1949), but non-directional analysis would detect no differences among the character states of taxa being compared. An accurate directional analysis would depend on an accurate reconstruction of ancestral character states. For example, if higher level nodes were estimated as the average of daughter taxon values, directional analyses would reveal no differences between primitive and derived forms.

and phylogenetic arguments these analyses can also make inferences about adaptations. Many of the newer techniques for non-directional analysis work at the level of making pairwise comparisons between closely related taxa as a way of assessing the number of times that some evolutionary relationship has evolved.

1.5 The comparative method and other ways of studying evolution

Evolutionary biologists use a variety of different methods for studying their subject. The results of comparative studies are often used to complement those from observations and experiments. The interpretation of results from comparative studies depends on the statistical test used, which is itself based on an implicit or explicit model of evolutionary change. In this section we examine briefly the relationship of comparative studies to other methods for solving problems in evolutionary biology. We start by distinguishing between two ways of modelling evolution (kinematic and optimality), both of which have their use in comparative studies. We then examine two case studies where the optimality class of model has taken a central role for integrating observational, comparative and experimental results.

There are two fundamentally different approaches which are in common use for studying evolutionary change and evolutionary equilibria. The first is the kinematic approach used by population geneticists (e.g. Lewontin 1974; Crow and Kimura 1970). Genotypes (or phenotypes) are specified with set frequencies, selective coefficients, migration rates, and mutation rates; and, perhaps, population sizes are specified in a spatially structured population. The model is then set to run for a specified number of generations, or equilibrium gene frequencies are sought. We shall use a kinematic model as our null model for independent character evolution in Chapter 4.

The second approach, favoured particularly by those studying behavioural ecology and animal mechanics, is optimality modelling (Maynard Smith 1978; Oster and Wilson 1978; Alexander 1982; Krebs and McCleery 1984; Stephens and Krebs 1986). Much of the rest of this chapter will be devoted to explaining what optimality models are, and how they are used in comparative tests. We stress from the outset that we are not advocating the unbridled use of optimality models, but pointing to their value as an aid for understanding many naturally complex systems, not least because of their ability to generate testable predictions.

The procedure for an optimality test is to specify an optimality criterion, to define alternative strategy sets, and to estimate the pay-offs for each strategy thus determining which is optimal under the conditions specified. The optimal strategy is the one predicted to occur. For example, foraging birds may have been selected to maximize the weight of food brought back to nestlings per unit time (optimality criterion). If there are alternative prey types consisting of large and small worms, birds might adopt one of three foraging strategies: gather all worms encountered, gather large worms only, or gather small worms only (alternative strategy sets). Different foraging strategies will result in different weights of worm brought back to the nestlings per unit time (different pay-offs), depending perhaps on the handling time for each type of prey, the weight of large versus small worms, and the time taken to travel between food patches and the nest. Birds will be expected to adopt that foraging strategy with the highest pay-off.

The reason for using optimality theory in evolution is that, subject to certain constraints, natural selection is expected to maximize Darwinian (or, more properly, inclusive) fitness, which is therefore the appropriate optimization criterion. A bird's beak might be engineered so that it provides an optimal tool for catching worms, and efficiency at catching worms in the short term may be a suitable optimization criterion as far as foraging is concerned. Birds which catch more worms leave more offspring. Optimal foraging theory considers a number of short-term optimization criteria, any one of which might ultimately maximize Darwinian fitness in a particular case. According to circumstance, maximizing the mass of worms provided per unit time, minimizing the amount of time taken to produce a set mass of worms, or minimizing the energy utilized while catching a fixed mass of worms might be an appropriate optimization criterion. For example, if foraging entails an increased risk of predation for either the adults or for the nestlings while the adults are away from the nest, time minimization might be more suitable than energy maximization. On the other hand, if predation is unimportant, a parent's Darwinian fitness might be limited by the rate at which it can feed its offspring.

One of the valuable facets of optimality modelling is that specified constraints are built into the system. We assumed above that a foraging bird was constrained to eat worms but not nuts, and that it took a set amount of time to handle a prey item or to travel from the nest to the foraging patch. Optimality models force us to make our constraints and our assumptions clear. In fact, one model's optimality solution may be another model's constraints. We can illustrate this point using the feeding bird example. One of the constraints concerned travel time. If larger birds could travel faster but need to use more energy for both maintenance and movement, we might reasonably seek the optimal body size for a bird that used a particular foraging strategy. Because the extra energy that a larger bird uses must be provided by food, our optimality criterion might be that body size which maximized the net rate of return of energy to the nest.

1.5.1 Implicit and explicit optimality models

Many of the best comparative tests, particularly those in animal mechanics, are based on explicit optimality models (Alexander 1968, 1982). However, preliminary comparative tests of adaptationist ideas are often based on *implicit* optimality models. We do not see this as a problem, but as a reasonable first step in a scientific investigation. For example, some years ago it was unclear to many primatologists whether sexual dimorphism in body mass among monkeys and apes resulted from selection for feeding niche differences between the sexes, as seemed likely for some birds (Selander 1972), or from sexual selection. If sexual selection is important, sexual dimorphism should be more pronounced among the more polygynous species, where one or a few males denies other males access to a group of breeding females. Polygyny places a premium on larger male body size because larger males are more likely to win fights. On the other hand, if feeding niche differentiation is more important, monogamous pairs living in a shared territory should be the most dimorphic. As it turned out, the sexual selection explanation made the correct prediction not only for primates (Clutton-Brock and Harvey 1977; Clutton-Brock *et al.* 1977) but also for several other groups of vertebrates including other orders of mammals, as well as some reptiles and amphibians (Alexander *et al.* 1979; Gittleman 1983; Shine 1978, 1979; Berry and Shine 1980). It took several years before more explicit optimality models were developed to predict or explain the degree of dimorphism for particular species. Sandell's (1989) optimality model for the evolution of sexual dimorphism in stoats (*Mustela erminea*) is a fine example. This, we believe, is an appropriate role for comparative studies of adaptation. They can usually demonstrate the generality of adaptive trends but, in the absence of suitable data, often they cannot test the fine details. We need to describe and explain both general trends and fine details if we are to achieve the real synthesis in evolutionary biology, which is to understand organic diversity. We shall return to this point later. Before doing that, we shall show that optimality models in comparative studies are often less explicit than may appear at first sight.

Explanations of comparative associations based on quantitatively defined functional relationships can mask unstated (and possibly unrealized) assumptions. Armstrong (1983), for example, claimed that the 0.75 exponent linking adult brain (A) to adult body mass (B) among mammals of different species ($A \propto B^{0.75}$) was a consequence of metabolic turnover constraining adult brain size: brains are energetically costly structures, and mammals have the largest brain mass that their metabolic turnover can maintain. Martin (1981, 1983) had both pre-empted Armstrong's explanation and cast doubt on it when he pointed out that the brain mass of birds and reptiles increases with the 0.56 power of adult body mass. Martin suggested that metabolic rate did indeed limit the evolution of brain mass, but through a different route: neonatal brain mass in mammals (N) was limited by the rate at which mothers could provide nutrition for their offspring. Maternal basal metabolic turnover (M) scales with the 0.75 power of body mass across species ($M \propto B^{0.75}$), and adult brain mass scales in direct proportion to neonatal brain mass ($A \propto N$). Martin argued, correctly, that the exponent linking the neonatal brain mass to the mother's body mass in mammals should be 0.75 ($N \propto B^{0.75}$). He also suggested that the 0.56 exponent in birds and reptiles was a consequence of their laying eggs. Mother's body mass constrains egg mass, and egg mass constrains neonatal brain mass, both according to a 0.75 exponent. As a consequence, neonatal brain mass should scale on mother's body mass to the $(0.75)^2$, which is about 0.56.

Martin's explanation is based on tests that involve juggling exponents, but that is no bad thing in itself because it may give biological insight. But why should the particular exponential relationships hold? If we take the relationship of metabolic rate to adult body mass as given, even if we do not yet know the cause, what are the supposed costs and benefits to a mother of providing more or fewer calories to her offspring's brain? Why should mothers belonging to related species of different sizes be able always to provide the same fixed proportion of their metabolic turnover to their offspring? Furthermore, even if it is plausible that mothers should supply a constant proportion of their metabolic turnover to their eggs, why should neonatal brain mass scale on egg mass with an exponent of 0.75? The egg must provide nutrient to nurture the young and its growing brain. It is therefore necessary to know the curves of neonatal mass and neonatal brain mass against time of incubation, and to compare these across eggs of different sizes, as a first step towards estimating how neonatal brain mass or neonatal mass might be expected to scale on egg mass.

As it happened, there was no need to develop Martin's argument because mothers belonging to species with high basal metabolic rates for their body size do not produce young with relatively large brains, as Martin's theory would predict (Pagel and Harvey 1988*b*, 1990). In other words, there are species deviations from the allometric relationship linking maternal body weight to metabolic turnover which do not correspond to deviations from the relationship linking maternal body weight to neonatal brain weight. If Martin's prediction had held, the next investigative step might have been to develop an explicitly formulated model, the assumptions of which could be tested, possibly using comparative data.

1.5.2 Symmorphosis as an explanation for scaling laws

Perhaps the most ambitious claim for optimality in evolution is symmorphosis. The idea is that each of the components of a physiological system should match the maximal requirements of the overall system. The system must perform at maximum capacity, but components are not likely to be over-designed because 'maintaining biological structures with their high turnover rates is costly' (Taylor and Weibel 1981, p. 3). Although Taylor and Weibel argued that symmorphosis may serve as a unifying principle for anatomy, Calder (1984) went further and suggested that it may provide a theoretical basis for scaling relationships. Weibel and Taylor (1981) predicted that the interspecific scaling of respiratory structures across animals of different size should match the scaling of maximal oxygen consumption. On the whole, the data went against their prediction. Garland and Huey (1987) took a slightly different approach and asked if species with high or low rates of maximal oxygen consumption for their size were those with high and low structural capacities to match (e.g. pulmonary diffusing capacity, mitochondrial volume densities, and capillaries per cross-sectional area of muscle). Again, the results tended to go against the idea of symmorphosis. Various explanations for the results are possible. One is that the components of the respiratory system have not been selected to function independently of other physiological and biochemical systems. Over-design for one system may constitute adequacy for another (Garland and Huey 1987). One of the heuristic contributions of the idea is that comparative results may reveal excess capacity for one function, and a legitimate question is now to seek the cause. For example, why are there more mitochondria in some species than appear to be needed by their respiratory systems working at maximum rate?

1.6 Testing adaptationist ideas using experiment, observation, and comparison

We have seen that comparative relationships can suggest adaptive scenarios which can then be tested and often falsified using other comparative relationships. What, then, can experiments achieve that comparisons cannot? One answer is that comparisons can provide an internally consistent story but, unlike experiments, they usually cannot distinguish cause from effect. (If ancestral character states were accurately recorded, prior evolution of one character state invariably followed by the evolution of a second character state may in fact distinguish cause and effect, as we saw in Section 1.4.2). Another advantage of experiments is that only one variable (the test variable) need differ among the various treatments, whereas with comparisons it is likely that many uncontrolled variables differ among taxa in addition to the variable of interest.

The complementarity of experimental and comparative approaches developed within the framework of optimality theory is nicely illustrated by work on sex ratio variation in the Hymenoptera and on foraging models and territory size among birds. These examples also demonstrate how optimality models can sometimes be used to provide quantitative predictions about comparative relationships.

1.6.1 The sex ratio

Fisher's (1930) argument for the equality of sex ratios was an early application of optimality theory to evolution. The optimality criterion is to leave the greatest number of grandchildren, and the strategy set is defined by the range of probabilities (zero to one) that any offspring produced will be female. Fisher assumed that male and female offspring cost different amounts, so that the total number of offspring produced by a female depended on the sex ratio, and he concluded that the evolutionarily stable investment strategy was to invest the same into both sexes. If two males cost the same to produce as one female, the evolutionarily stable sex ratio would be two males to one female. Several assumptions were built into Fisher's model and, as they are changed, so does the optimal investment strategy (Hamilton 1967; Charnov 1982).

Hamilton (1967) suggested that producing sons can often result in diminishing returns. For example, if a mother's male offspring competed among only themselves for exclusive mating access to females (Hamilton termed this 'local mate competition'), then one son would produce as many grandchildren for her as would any number of additional sons. The mother would leave more grandchildren if she produced one son and invested the other resources in daughters. Furthermore, if mating between sibs is common in the population, there is a second advantage for a mother producing more daughters: they provide additional mating opportunities for her sons (Taylor 1981). Several models predict the equilibrium sex ratio under different degrees of local mate competition and inbreeding (see Harvey 1985).

Hymenoptera provide good material to test these ideas because of the mother's ability to determine the sex of each offspring by deciding whether or not to fertilize an egg (Hamilton 1967): fertilized eggs become daughters and unfertilized (haploid) eggs develop into sons. Differences in population structure among species, associated with differences in local mate competition and degree of inbreeding, have provided useful material for comparative tests.

Sex ratios might be expected to be biased towards females when the brood mates before dispersing, because in such species both local mate competition and inbreeding are likely to be common. Scolytid bark beetles are of particular interest here (Charnov 1982) because some species mate before and others after dispersal. As theory predicts, those species which mate before dispersal have female-biased sex ratios, whereas those which mate after dispersal have sex ratios near equality. Similarly, the presence of winged males in fig wasps is an indicator of how much mating takes place after dispersal, and there is a positive relationship between the absence of winged males and the proportion of females (Hamilton 1979). In another notable comparative study, Waage (1982) examined sex ratio variation among species of Scelionid wasps which parasitize the eggs of other insects. One parasitoid egg is laid per host egg. Waage argued that those host species which laid small clutches would be exploited by a single parasitoid at most, but when host clutch size was large several Scelionid females might parasitize a single host clutch. Local mate competition should, therefore, be more intense when host clutches are small, and we should expect a positive correlation between the proportion of males per parasitoid clutch and host clutch size. Waage's data demonstrate the predicted correlation (Fig. 1.10), with the exception that when host species laid single eggs the parasitoids did not produce a heavily female-biased sex ratio, which was to be expected because the singletons produced would have to disperse in order to find a mate.

Although the cross-species comparative data can be nicely explained by the models, it is always possible that some other component of the biology of the species in the sample is responsible for the sex ratio differences. Experimental studies performed *within* species can control for such variables. For example, laboratory experiments on the wasp *Nasonia*

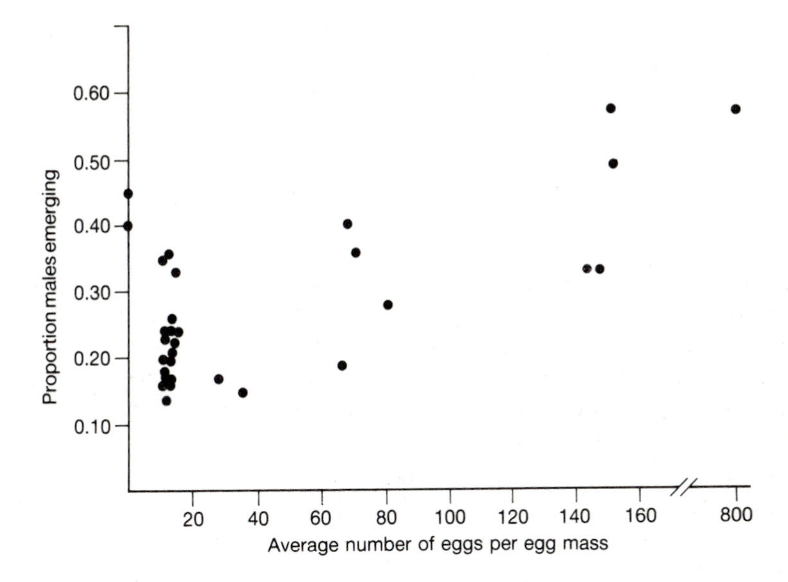

Fig. 1.10. The proportion of males emerging from Scelionid wasp clutches plotted against host egg mass size. Each point is a different species. Sex ratio theory, invoking local mate competition, predicts both the overall positive correlation and the observed exception that host species laying single-egg clutches have parasitoids with a sex ratio near 50:50. (After Waage 1982).

vitripennis showed that the sex ratio of emerging adults shifted from a strong female bias towards equality as more parasites were given the opportunity to lay their eggs on groups of host pupae (Walker 1967). This is to be expected because both local mate competition and inbreeding should be reduced when more broods are laid per host and when wasps are emerging simultaneously from nearby hosts. Subsequent work by Werren (1980, 1983) demonstrated that super-parasitism and host clumping resulted in sex ratio shifts, but that sex-specific mortality among parasites was not involved. Furthermore, the sex ratios in the smaller brood laid by super-parasitic second females accorded with theoretical predictions based on brood sizes and the proportion of males laid by the first female (see Fig. 1.11). (Orzack's (1986) experimental results from the same species found a poor fit with theory which, he suggests, is due to constraints on the wasps' ability to detect previous oviposition and to produce an exact sex ratio.) Other experimental studies also provide excellent tests of the theory (e.g. Herre 1985, 1987).

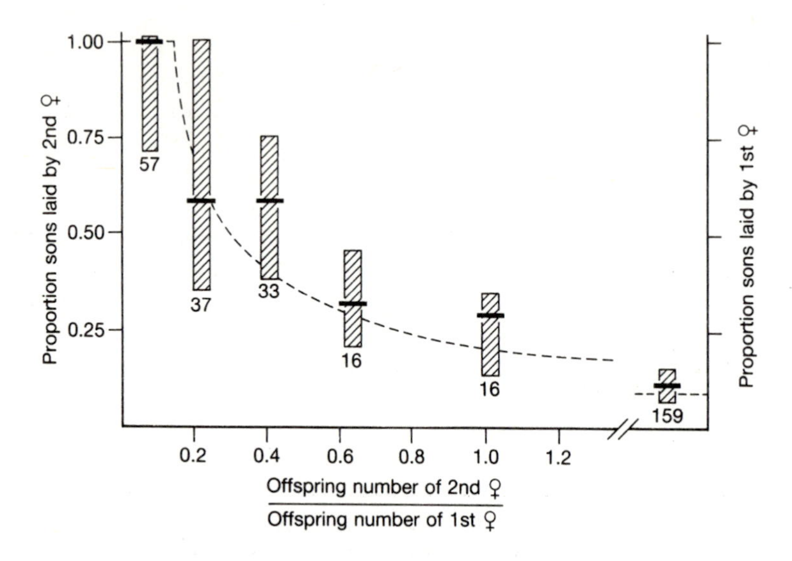

Fig. 1.11. Super-parasitic female wasps of the species *Nasonia vitripennis* lay clutches with higher proportions of male eggs. Data from 159 super-parasitized hosts show that, as predicted by sex ratio theory (dotted line), the super-parasitic female decreases the proportion of sons laid as a function of her brood size relative to that of the first female. (After Charnov 1982).

1.6.2 Home range and and territory size

A similar but less complete story can be told for the development of our understanding of the determinants of variation in territory and home range sizes among vertebrates. Preliminary comparative studies in the 1950s, 1960s, and 1970s examined the ways in which species' territory and home range sizes change with body size and diet (Hutchinson and MacArthur 1959; McNab 1963; Schoener 1968; Milton and May 1976). A reasonable question to ask at that time was whether larger animals need larger territories to satisfy their larger metabolic needs. As predicted, territory size did increase with body size and, furthermore, species living on more sparsely distributed food resources also had larger territories.

The next step was to ask whether territory size increased with body size in a quantitatively sensible way. Because, as we have seen, metabolic rate increases with body weight raised to the 0.75 power, presumably the minimum size, continuously productive territory (energy produced per unit time) for animals with similar diets living in similar habitats would also be expected to increase with the 0.75 power of body mass (*contra* Lindstedt *et al.* 1986). The data did not accord with that expectation: territory size in a variety of taxa increased with body size with an exponent appreciably greater than the metabolic exponent (e.g. Harvey and Clutton-Brock 1981; Gittleman and Harvey 1982; Mace and Harvey 1983; Lindstedt *et al.* 1986).

Several possible causes for this discrepancy seemed likely. First, suitable habitat is not continuous, and larger-bodied species must take in a disproportionate area of unsuitable habitat. Second, the acceptable food spectrum may change with body size (Schoener 1983). For example, individuals supplying food to a nest in the centre of a territory might have evolved an optimal foraging strategy resulting in the selection of only the larger food items at an increased foraging distance from their nest. Larger bodied species might become increasingly selective at greater distances from the nest, possibly because of intruder pressure at the nest. It is both unnecessary and impractical to perform detailed field observations and food manipulation studies on more than a few species in order to determine the likely reasons for the comparative relationship. A series of optimality models has been developed, based on a variety of optimization criteria, which predict different relationships between territory size and body size (Schoener 1983). Also, the likely optimization criteria are becoming better known as a consequence of carefully controlled field and laboratory experiments (Davies and Houston 1984; Stephens and Krebs 1986). The integration of foraging theory with comparative studies has been a long time coming, in part because it has proved difficult to paramaterize simply the cost and benefit curves for foraging animals.

1.7 Defining characters and environments

We have described the comparative biology that concerns us here as belonging to the 'the guild school'. That may have seemed an unfortunate denomination because ecological guilds are rather fuzzy objects (but see Adams 1985). Originally, Root (1967, p. 335) defined a guild as 'a group of species that exploit the same class of environmental resources in a similar way'. Just as Root's concept has been of enormous heuristic value in ecology, so a catholic interpretation of his definition captures the essence of character states and environments that are used in comparative analyses. For example, winged animals exploit air for flight and animals with eyes use light for vision. Both wings and eyes have evolved on separate occasions in different lineages. If we were interested in adaptations for flight, we should compare the guild of flying animals (which would contain most birds, bats, and beetles) with each other and with the guild of their flightless relatives (such as crocodiles, cetaceans, and collembola). However, if our interest was adaptations for a carnivorous diet, our guild structures would naturally change.

Character states or environments may need to be redefined as we learn more about the problem we are tackling, particularly about how our study organisms interact with their environments. After all, if we were interested in coloration among insects as an adaptation to avoid predation by reptiles and we found that reptiles do not have colour vision, we should translate our colour scores to a monochromatic scale. To illustrate the major problems associated with the definition of character states and environments, we shall take two further examples in a little more detail. The first is beak morphology in birds and the second is rates of genetic recombination in mammals.

Our first example concerns the need for careful definition of environments. Hawks, shrikes and Australian shrike-tits have very similar beaks. A comparative analysis suggests that long curved beaks have evolved independently in each taxonomic group. Beaks are used for feeding, so we might assume that the three groups have similar diets. We should be wrong. Hawks and shrikes are predators of small vertebrates and large insects, whereas shrike-tits feed on small insects. However, to get at the small insects, shrike-tits need to rip the bark from trees. Although their diets are not the same (Simpson 1978, p. 218), 'in all three groups the function of the beak is to seize and rend and the fact that different things are seized and rent is irrelevant. The functions are the same in all three, and an inference from comparison of similar selection pressure and similar featural response is valid'.

The procedure we have described for the investigation of factors influencing beak shapes has been labelled 'progressive ad hoc optimization' by Lewontin (1979). If theory does not explain the data (diet is not correlated with beak shape), then adjust the theory a little until it does (feeding mode is correlated with beak shape). Lewontin thinks that this is an unsatisfactory way to do science. We see it as a useful way of establishing the truth.

Our second example illustrates how characters may need to be reassessed in the light of comparative evidence. Why do rates of genetic recombination vary among mammals? Interspecific comparisons reveal fairly poor relationships between recombination rates and factors that might be expected to influence them (Burt and Bell 1987). For example, higher host recombination rates may be favoured when parasites have several generations per host generation to evolve pathogenic strains. Other things being equal, higher recombination rates might then be expected in host species with longer generation times. Only about 22 per cent of the interspecific variance in recombination rates is accounted for by generation length, estimated by age at maturity. But recombination rates depend on both chromosome number and chiasma frequency. The percentage of variance in chromosome number and in a measure of chiasma frequency accounted for by age at maturity are about 1 and 77 per cent respectively.

In fact, it has proved very difficult to find any good correlates of chromosome number, despite years of effort (Williams 1966; Bell 1982); perhaps the number of chromosomes is a character that does not respond readily to selection? In any event, the constituents of recombination rate should be treated separately in comparative analyses. Biologists were not to know this in advance. The most appropriate atomization of characters is not always obvious (Thompson 1917, p. 713; Gould and Lewontin 1979). The recombination rate example also illustrates how in our search for the adaptive significance of character states, we often seek associations between pairs of characters rather than between characters and environments.

Just as we take a liberal interpretation with the word character, so we do with the concept of phenotype. For the purposes of this book both character and phenotype can refer to any morphological or behavioural trait that differs among species. Very often we do not know the extent to which interspecific differences are caused by environmental or genetic differences, and we usually have little information about their ontogeny. Most comparisons are among adults, although from time to time we shall refer to differences among juveniles. Our approach is pragmatic. We want to reveal new patterns and to test interpretations of old ones. Such studies heighten our appreciation of the importance of both development and genetics in the study of evolution. Comparative studies reveal crucial unexpected differences or lack of differences among taxa that point to the importance of developmental systems and genetic constraints (by which we mean lack of suitable genetic variance on which selection can be effective). We have no doubt that, as they have in the past, comparative studies in the future will reveal new cases of heterochrony and neoteny (Gould 1977). As we shall discuss in Chapter 2, the absence of some comparative trends in some taxa is undoubtedly a result of insufficient suitable genetic variance on which selection can act—and differences in comparative trends between taxa are often the result of different phenotypic responses to similar selective pressures. The extent to which those different responses might be viewed in terms of proximity to alternative adaptive peaks in the sense of Sewall Wright (1932) is open to both question and test (Ridley 1983*a*).

1.8 Summary

When similar character states evolve independently in similar environments, it is natural to ask how they adapt their bearers to survive and reproduce in those environments. However, similarity among closely related species provides evidence for identity by descent from common ancestors. Modern comparative methods attempt to distinguish independent evolutionary origins of character states from cases of identity by descent, even though both may have an adaptive basis: similarity among closely related species may be selectively maintained. When attempting to interpret comparative evidence, it is also important to distinguish the selective forces responsible for the origin of character states from those responsible for their maintenance in contemporary populations. Many comparative tests are based on optimality models, and complement the testing of adaptationist ideas by experimentation, which is often impractical.

2

Why worry about phylogeny?

'Comparative biologists may understandably feel frustrated upon being told that they need to know the phylogenies of their groups in great detail, when this is not something they had much interest in knowing. Nevertheless phylogenies are fundamental to comparative biology; there is no doing it without taking them into account' (Felsenstein 1985*a*, p. 14).

'Ought we, for instance, to begin by discussing each separate species—man, lion, ox, and the like—taking each kind in hand independently of the rest, or ought we rather to deal first with the attributes which they have in common in virtue of some common element of their nature, and proceed from this as a basis for the consideration of them separately?' (Aristotle, *De partibus animalium*).

2.1 Introduction

Living organisms can tell us a lot about their evolutionary history. Indeed, our estimates of phylogenies would be much the same in the absence of a fossil record. This chapter explains why an assessment of phylogenetic relationships is a prerequisite for a successful comparative analysis. In particular, closely related species share many similarities in addition to those of relevance to any particular comparative question. Such similarities can confound comparative studies. If we had a sample of bird and mammal species, for example, and wanted to know why some species have feathers, we might notice that the feathered species lay eggs and have beaks while those species with teeth and fur produce live young. As far as we know, these differences are not adaptively related, but if each species in our sample was used as an independent point for statistical analysis, we should find strong associations between having feathers and beaks and laying eggs. But a phylogenetic reconstruction of this case would reveal that the characters in question had each evolved just once in these groups. This is why we must worry about phylogeny: phylogenies help us to identify *independent* evolutionary events, and it is independent events that statistical tests rely on. This theme will recur throughout.

After describing how a knowledge of phylogenetic relationships can be used in comparative analyses, we shall discuss the biological foundations for phylogenetic history being retained in contemporary phenotypes. This will lead to an examination of the biological reasons why closely related species are so similar to each other. In Chapters 4 and 5 we provide a statistical account of phylogenetic similarity to complement the biological perspective given here.

2.2 Correlation, causation, phylogeny, and confounding variables

The correct use of phylogenetic information can help distinguish cause from effect in comparative relationships, and also eliminate many potential confounding or third variable explanations. Furthermore, if the same correlations between character states can be shown to exist in several independently evolving lineages, this means the traits have tended to evolve in a correlated fashion and explanations associated with phylogenetic history are unlikely to apply.

2.2.1 Distinguishing cause and effect

Comparative studies relating either phenotype to environment or phenotype to phenotype among contemporary species are inevitably based on correlational evidence. But, as common sense and most elementary statistical textbooks tell us, correlation is not causation. Consider two characters that can each exist in one of two states. For example, butterfly larvae may be palatable ($P+$) or unpalatable ($P-$), and they may be solitary ($S+$) or gregarious ($S-$). Suppose we find that larvae from palatable species tend to be solitary, so that character state $P+$ is associated with character state $S+$, then by paying attention to phylogenetic history, we can begin to unravel causation. The association between $P+$ and $S+$ may have arisen because (1) $P+$ causes $S+$; because (2) $S+$ causes $P+$; or because (3) both $P+$ and $S+$ are caused by some third variable. A good phylogenetic tree with specified ancestral character states allows us to distinguish directions of causality: if $P+$ always appears before $S+$ in a phylogeny and we can rule out other causal influences, then the direction of causality is established (Ridley 1983*a*; Chew and Robbins 1984).[4] This approach has been used to investigate, for example, displays and the evolution of polygyny in birds (Winterbottom 1929), the evolution of patterns of parental care in fish (Gittleman 1981), and the evolution of gregariousness and aposematic coloration in lepidopterans (Sillén-Tullberg 1988).

---

[4] We are assuming here that there are no time lags in the system. For example, a third character could change, bringing about an immediate response of $P+$ and then a lagged response of $S+$.

2.2.2 Removing the influence of confounding variables

The problem of confounding variables is likely to be reduced but not necessarily eliminated by searching for the same relationships in different lineages (Clutton-Brock and Harvey 1979). On the whole, closely related species are more similar than distantly related species in morphology, behaviour, and ecology. If closely related species share a character state, then the chances are that they will share a whole lot more too. Consider a comparison among species belonging to 2 distantly related genera, say 10 species of *Peromyscus* mice and 10 species of *Drosophila* fruit flies. Differences among the mice or among the flies are likely to be swamped by differences between the two groups. If species are treated as independent points for analysis, we would find significant associations and correlations between almost any pair of characters we examined: diet, body weight, leg number, presence or absence of wings, clutch size, lifespan and so on. Such relationships are unlikely to be informative in our search for the adaptive significance of cross-taxonomic variation.

However, rather than making comparisons between genera, we might instead look within genera. (For didactic purposes, we are assuming that our phylogeny is incomplete and the true relationships among species within genera are unknown, which is not actually true for *Peromyscus* and *Drosophila*.) Comparisons within genera can, of course, deal only with characters that actually vary within genera, like diet, body weight, clutch size and lifespan. However, when they can be made (e.g. Read 1987), such comparisons are particularly informative because they automatically hold constant all the variables that are shared by congeners (see Møller and Birkhead in Harvey 1991).

2.2.3 When patterns differ among taxa

Cross-species comparisons can mask interesting patterns in the data (Clutton-Brock and Harvey 1984). It is frequently the case that correlates of variation in some taxa differ from those in other taxa, and we shall meet examples in the next section. However, on occasion it seems that relationships between variables at one taxonomic *level* are not the same as those at another: the same patterns may be repeated in different taxa, but vary according to the rank of the taxon being considered! An extreme example is given by Huey (1987): the relationship between variability in body temperature and thermal performance breadth (the range of body temperatures through which lizards' sprint speeds exceed some arbitrary criterion) is positive within a number of genera but negative across genera. Another example (which we shall question in Chapter 6), is the suggestion that brain mass increases less for a given change in body mass among adults of more closely related species of vertebrates than it does across adults of more distantly related species (Lande 1979; Martin and Harvey 1985; Pagel and Harvey 1989*a*). For example, the exponent relating brain to body mass among species within genera is often said to lie between 0.2 and 0.4 (Gould 1975; Lande 1979), whereas among species from different orders it is typically between 0.55 and 0.75 (Martin 1983; Harvey and Bennett 1983).

The patterns in the two examples described above would have been missed if phylogenetic relationships were not studied. If such phylogenetically related patterns are commonplace, we must be very careful to take full account of phylogenetic relationships in comparative studies. One pattern that we can be sure *is* commonplace is similarity among closely related species, but the processes responsible for that similarity are less often considered.

2.3 Three reasons why phylogenetically related species are similar

How has evolutionary history come to be recorded in contemporary phenotypes? At least three processes have been involved which we term *phylogenetic niche conservatism, phylogenetic time lags*, and *different adaptive responses*. We consider each in turn below, and illustrate them as problems that comparative analyses must face by drawing an analogy between comparative and experimental studies in Box 2.1. Comparing phenotypes of extant species living in known environments is like analysing the results of a temporally nested experiment when records of early treatments and responses to them have been lost.

2.3.1 Phylogenetic niche conservatism

The first process leading to similarity among closely related species is adaptive, and is a consequence of vacant niches having been invaded by those available species that were best suited to occupy them. For example, if a new area of coniferous forest appeared adjacent to grassland, lakes, and deciduous forest, it is more likely that it would have been invaded by species of, say, birds and insects from the deciduous forest than from the other two habitats. Fish would not stand a chance in the new terrestrial habitat, and birds adapted to grassland would probably lose in competition with those from deciduous forests. Through time, the birds from the deciduous forest would, no doubt, become better adapted to the new coniferous habitat and diverge phenotypically from their ancestors. But they would still be birds, and they would retain the lifestyle of birds. They would not become insects, partly because insects would be occupying the insect niche. This is the principle of phylogenetic niche conservatism: past and present phenotypes of a lineage are likely to have occupied similar environments. This is a purely adaptationist reason for why phenotypically similar species are likely to be close phylogenetic relatives (Grafen 1989).

Box 2.1. An experimental analogue for comparative studies

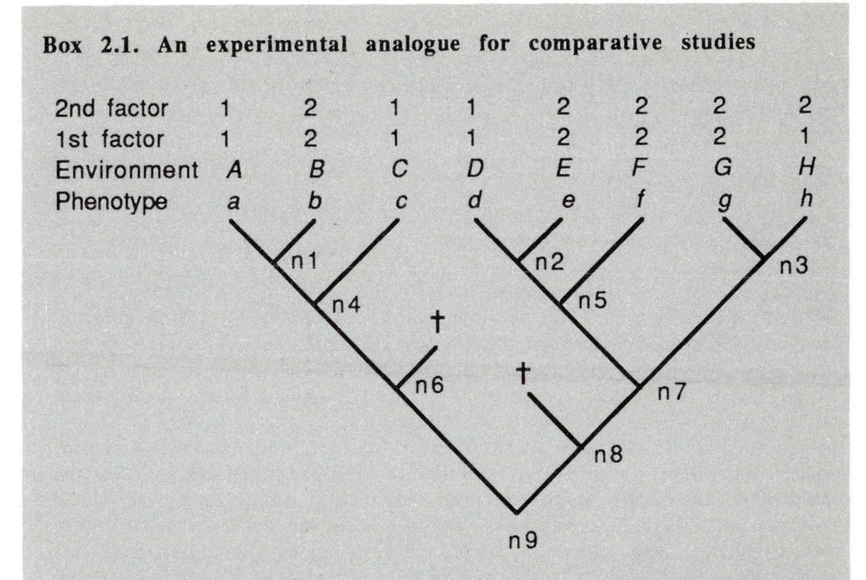

An analogy between (1) phenotypic evolution under a range of environments, and (2) response of individuals to a nested sequence of experimental treatments.

1. Contemporary phenotypes *a* to *h* are found in environments *A* to *H*. The ancestral species marked at nodes n1 to n9 existed when lineages divided. Some lineages went extinct (†). Because historical information on phenotypes and environments is lacking, both must be reconstructed.

2. The experimental analogue begins by subjecting a large sample of individuals to the same treatment at time n9. The individuals are then divided sequentially into groups and subgroups repeatedly but after varying intervals of time. Between divisions, the individuals in each group or subgroup are subjected to the same sequence of treatments which differs from that received by other individuals in other groups. The experimentalist keeps no historical record of the treatments received by the different individuals. Because historical information on treatments and responses is lacking, both must be reconstructed.

Knowing the structure of the tree (the sequence in which lineages split), we can use contemporary phenotypes and environments (treatments) to help in that reconstruction (this will be the topic of Chapter 3). Further, because extant phenotypes contain information about their evolutionary history, the comparative method must identify phylogenetic correlates in order to distinguish them from the effect of current environments on current phenotypes. Bearing in mind the analogy between phylogeny and a nested experiment, the effect of history on phenotype can be summarized through five generalizations.

1. Each environment consists of many selective factors or, equivalently, each treatment consists of many factors. Levels of two factors are shown in the figure, but contemporary and historical environments differ on many other non-independent factors. Factors 1 and 2 might be measures of temperature (hot, cold) and humidity (wet, dry), but other factors such as whether food sources are clumped or not are ignored.

2. Factors tend to covary across environments or treatments. For example, the level of factor 1 is a good predictor of the level of factor 2. (This results in the problem of *confounding variables* discussed earlier in this chapter (Section 2.2): a relationship between variation in a phenotypic character and an environmental factor may be caused by another, possibly unrecorded, factor).

3. Past environments or treatments for a lineage tend to be similar to present environments. For example, n3 is likely to have had level 2 of factor 2 because phenotype *g* does. As far as the experiment is concerned, this would mean that the past treatments on a sample were similar to subsequent treatments. For the evolutionary analogue, it means that organisms tend to occupy similar habitats to those occupied by their ancestors. (This may be because, when new ecological niches open up, those species inhabiting the most similar existing environments are best adapted to invade the new ones. This is *phylogenetic niche conservatism:* closely related species may be phenotypically similar to each other in part because they are likely to inhabit similar environments *for historical reasons*. Unrecorded factors in the environments of close relatives are likely to be more similar to each other than are unrecorded factors in the environments of distant relatives.)

4. Past environments or treatments may leave their marks on contemporary phenotypes. For example, the environment for n7 may have been influential in moulding phenotype *d*. Furthermore, random changes resulting from neutral evolution by genetic drift or some change in experimental phenotypes that did not result from specified treatments may be retained. (This is the problem of *phylogenetic time lags*: it is possible that a character state observed on a contemporary phenotype represents an adaptation to factor(s) in some past environment. The non-selective version of time lags is genetic drift or neutral evolution: species that have shared longer periods of phylogenetic history may still have similarities that accumulated as a consequence of genetic drift. We shall illustrate this process in Chapter 5, Section 5.2 and Fig. 5.4.)

5. The response of a lineage to a treatment (environment) depends, to some extent, on its phenotype when it entered that regime. For example, phenotypes *a* and *d* have similar levels of factors 1 and 2 but, if they differed phenotypically at n1 and n2, they might respond differently to factors 1 and 2 and move towards different equilibrium phenotypes. (*Different adaptive responses* to similar selective forces are a problem facing many comparative studies: two types of organisms may have responded to the same environment in different ways.)

There are two components to the concept of phylogenetic niche conservatism: (1) the species most likely to invade a vacant niche is the one in an adjacent environment that occupies the most similar niche; and (2) species are restrained from moving into new niches in large part because those niches are already occupied by other taxa that are well adapted to the niches and are better competitors for the limiting resources. Evidence for the general importance of phylogenetic niche conservatism may actually be provided by the sporadic occurrence of adaptive radiations: the absence of competitors that occupied adjacent niches in ancestral environments often provides the clue to understanding an adaptive radiation that followed a species' invasion of a new habitat. Darwin's finches on the Galapágos Islands, from which other land birds were absent, provides a typical example (Grant 1986).

2.3.2 Phylogenetic time lags

The second historical process involved in phenotypic diversity is phylogenetic time lags. A trait could have evolved in a common ancestor of several extant species, either by natural selection or by genetic drift. The trait will ultimately be lost provided that (1) suitable genetic variance is present or arises by mutation, and the costs of retaining the trait are not outweighed by its benefits in the contemporary environment, or if (2) pleiotropic gene effects are involved so that selected changes in another character lead to reduction in the focal character.

Phylogenetic time lags do not provide as important a problem to comparative studies as we might think. Related species may express traits that are of little or no adaptive significance, and which certainly do not serve the function they had in the past. The vermiform appendix of humans and the remnants of hip bones in cetaceans are cases in point. Our aim is to explain taxonomic diversity, and such organs usually vary in their state of development among higher-level taxa. If we look across mammals, we find that herbivores have a large vermiform appendix (the caecum) whereas carnivores do not, and terrestrial mammals have hips whereas whales do not. In such cases, the comparison can point to the functional significance of the trait of interest. The important point is that the correct comparison must be made. If we find an apparently functionless trait, a search for related taxa with the same trait differently developed may help to explain why the trait is there. It is not always necessary to seek comparison with distant relatives. For example, the flight motor neurons of flightless grasshoppers, which have been inherited from flying ancestors, are now smaller than the equivalent neurons of flying grasshoppers (Dumont and Robertson 1986).

Intimate knowledge of the biology of a species would be necessary to demonstrate that a trait is, in fact, of no functional significance. There are many examples of traits taking on different functions (Gould and Vrba 1982). This means that if some members of a taxon have a trait that can be shown to serve a particular function, it does not necessarily follow that the same trait serves the same function in other members of the taxon. For example, on occasion the kea *Nestor notabilis*, a New Zealand parrot, uses its beak to rip through the skin of dead sheep and feed on the fat beneath. Closely related parrots use their kea-like beaks to feed on seeds and fruit (Futuyma 1979). As we shall see below, different selective forces may go on to mould the same character.

2.3.3 Different adaptive responses

The third way in which history is recorded in contemporary species is through different adaptive responses. The methods used in later chapters of this book seek similar adaptive responses to similar selective pressures, recognized as instances of parallel or convergent evolutionary change. However, one important limit to correlated evolutionary change concerns the phenotypic similarity of the species being compared. Animals of similar phenotype are likely to evolve similar responses to the same selective force, whereas different phenotypes may respond differently. For example, in response to a threat from predators, distasteful animals might evolve warning coloration while palatable species would become cryptic (Harvey *et al.* 1982; Guilford 1985). In a similar vein, when faced with the threat of predation, large animals may stand and fight thus being selected for even larger size, whereas small animals may escape down burrows and as a consequence are selected to be even smaller (Edmunds 1974; Simms 1979; Ralls and Harvey 1985). Such differences can set phylogenetic limits to the generality of many comparative trends.

Adaptive responses lie behind many historical explanations of biological diversity just as they are at the root of many micro-evolutionary processes. Distantly related species occupying similar niches may remain phenotypically dissimilar, while closely related species may show parallel or convergent evolution. As we shall see below, the concept of different evolutionary responses to similar selective forces unites several biological perspectives on the problem of why closely related contemporary species tend to be phenotypically similar.

Different characters respond to similar forces

In order to browse from trees, giraffes have evolved long necks while some other species of mammals have evolved the ability to climb trees. The adaptive route taken may be influenced by phylogenetic history. It may also constrain or otherwise influence the subsequent direction of evolutionary change. For example, in response to increased predator pressure, terrestrial browsers can evolve even larger body sizes, but this route may be closed to arboreal browsers which must often be small enough to feed from terminal twigs without breaking them (Clutton-Brock and Harvey 1983). They may therefore evolve a nocturnal lifestyle with accompanying changes in their visual apparatus. This is essentially the phenomenon of an adaptive landscape having separate adaptive peaks (Wright 1932). There are many well-documented cases of different characters having responded to similar selective forces. Simpson (1967) cites the case of carnivorous mammals in which different teeth in the lower jaw have become adapted for meat shearing: the fourth premolar in the wolf *Canis*, the first molar in *Oxyaena*, and the second molar in *Hyaenodon*.

As Gould and Lewontin (1979) have emphasized, selection uses whatever variation is available, and if the variation is heritable a character will evolve. For example, male bovids have strong horns for combat over mates, whereas cervids use antlers. Cross-species comparisons within each of those families of mammals demonstrate expected relationships between the size of these costly structures and the extent to which the species is polygynous (Clutton-Brock *et al.* 1980; Packer 1983). However, comparisons of horn size and antler size with measures of polygyny across all mammals would not reveal the strong patterns found within the two families. Indeed, males of polygynous species in at least one order, the primates, have developed neither horns nor antlers, but enlarged canines (Leutenegger and Kelley 1977; Harvey *et al.* 1978).

The evolution of horns, antlers, teeth, and other weapons has been accompanied by both the evolution of different fighting strategies and defence of appropriate areas of the body. As a consequence, species differences in the location of dermal shields and areas of thickened skin only make sense in the context of differences in fighting behaviour and weaponry (Jarman 1988). Again, one adaptive route influences the direction of subsequent evolutionary change, a topic we shall need to return to below.

Studies of structural antigenic and genetic changes in haemagglutinin which accompany adaptation of an influenza virus to being cultured in hens' eggs provide a good example of alternative genetic mutants appearing to provide similar adaptive responses. When genetically identical isolates of the A(H1N1) virus were propagated in eggs, one of three antigenically distinct variants not present in the inoculum soon appeared in each egg and spread to fixation. Identical antigenic variants were not always genetically identical, although in every case HA1 amino acid substitutions were located in the vicinity of the receptor binding site (Robertson *et al.* 1987).

Different forces mould the same character

Just as different characters may respond to meet the same selective pressure, so the same character may respond to different selective pressures. A careful redefinition of the function of character states can sometimes reveal unrecognized similarities, as we saw with the beaks of hawks, shrikes, and shrike-tits in Chapter 1. But this will not always be the case. The forward pair of pentadactyl limbs in mammals provides an obvious example (Darwin 1859, p. 434): 'What can be more curious than that the hand of a man, formed for grasping, that of a mole for digging, the leg of the horse, the paddle of the porpoise, and the wing of the bat, should all be constructed on the same pattern and should include the same bones, in the same relative positions?'.

Different degrees of response

Different taxonomic groups may even produce different degrees of response to similar selective pressures. For example, genetic variance allowing the evolution of a particular adaptation may occur in one taxon but not another. Compare sex determining mechanisms in the hymenoptera and fruit flies. Hymenopterans have haplo-diploid sex determination with fertilized eggs developing into females and unfertilized eggs into males. Under conditions of regular inbreeding or local mate competition, the optimum strategy for a mother is to produce broods with an excess of daughters (see Section 1.6.1). Hymenopterans adapt their sex ratios accordingly (Hamilton 1967), as mothers can choose whether to fertilize any particular egg as it is laid (e.g. Gerber and Klostermeyer 1970). However, like mammals, *Drosophila* has chromosomal sex determination with XX individuals developing into females and XY into males, and it may be that here strongly biased sex ratios cannot easily evolve (Maynard Smith 1980). For example, attempts to bias the sex ratio by selection experiments have usually failed (e.g. Toro 1981; Toro and Charlesworth 1982), and when adaptive sex ratio variation can be detected among species with chromosomal sex determination, it is not so extreme as among the hymenoptera and other haplo-diploid taxa (Clutton-Brock 1986; Clutton-Brock and Iason 1986; Bull and Charnov 1988).

We do not need to compare such phylogenetically distant relatives as those from different orders or classes to find different degrees of response to selection. For example, the extent to which thermal sensitivity of sprint speeds evolves genetically to match activity body temperatures seems to differ among iguanid genera (van Berkum 1986). At a lower taxonomic level still, geographic variation in morphometric measures of a single species, the pocket mouse *Perognathus goldmani*, is more closely allied with phylogenetic relatedness among karyotypically defined races than with those environmental variables which are 'standardly employed in attempts to determine factors underlying patterns of geographic variation in morphology' (Straney and Patton 1980, p. 896). But as Straney and Patton stress, and as is usually the case in such fine-grained comparisons, the similarities may result from unmeasured selective forces that are more similar in the niches occupied by the more closely related races. Without a detailed genetic analysis, the results of controlled selection experiments, and appropriate observational studies of animals in the field, it is not possible to determine the cause of these and other similar patterns (see also Grant 1986, p. 183). Nevertheless, the probability that even different genetic races of the same species will not respond in precisely the same way to the same selective forces should always be borne in mind when interpreting imperfect comparative relationships.

Key innovations and developmental constraints

The topic of different responses to the same selective force subsumes two well-discussed biological perspectives on evolutionary change. The first is the concept of key innovations, and the other is developmental constraints. We mention these here, albeit briefly, because they have generated their own fairly voluminous literatures that bear on our theme. From time to time, a change occurs in a phenotype that allows a new range of viable variants to evolve. As a consequence, a new adaptive radiation can arise, either because the derived forms containing the new characteristic displace old forms from their niches or because the new characteristic is associated with higher rates of speciation or cladogenesis. Futuyma (1986, p. 439) provides a useful review of these areas and puts the matter in a nutshell when he writes that '(developmentally) integrated systems are likely to display a limited, recurring repertoire of variations, giving rise therefore to parallel evolution and to atavistic variants that reveal in a recapitulatory way the ancestral foundations of the developmental program'. A new mutation producing a key innovation may release the ancestral phenotype from some developmental constraint, thus setting the scene for a new adaptive radiation.

In a careful analysis of the pharyngeal bones and associated musculature of cichlid fishes and their relatives, Liem (1973, 1980) has probably pinpointed such a key innovation. Cichlids have undergone adaptive radiations in African lakes despite competition from species belonging to several other fish families. The success of cichlids results in large part from an evolved diversity of feeding mechanisms. In relatives of cichlids, the pharyngeal bones hold prey items but cannot manipulate them. A small shift in position of a single muscle attachment in a common ancestor of cichlids, possibly caused 'by a very simple change in ontogenetic mechanism' (Liem 1973, p. 439), may have reversed the muscle's function from one of abduction to adduction of the lower pharyngeal jaw. This was then followed by secondary changes that led to the pharyngeal bones being able to manipulate as well as to hold prey items. As a consequence, the premaxillary and mandibular jaws which had previously been constrained to serve the dual functions of collecting and manipulating food were freed to evolve along new routes that did not involve food manipulation. Furthermore, Liem (1980) goes on to show how an anatomical decoupling of the maxilla and premaxilla in cichlid fishes has allowed greater flexibility of potential feeding movements. For example, cichlids have five ways to protrude the upper jaw whereas their close relatives have just one.

Lauder (1981) has focused on how we might demonstrate that particular derived morphological features will trigger the evolution of phylogenetic lineages with parallel patterns of morphological change (transformations), resulting in arrays of terminal taxa with similar phenotypic ranges (relations). He argues that so-called transformations and relations may be demonstrated to be repeatable by showing how the same transformations and relations have occurred in different monophyletic groups, each containing the same key morphological feature (See Fig. 2.1). The hypothesis that the key feature is responsible for the historical consequences can be tested by comparing the transformations and relations among monophyletic taxa, some of which contain the key feature and others of which do not.

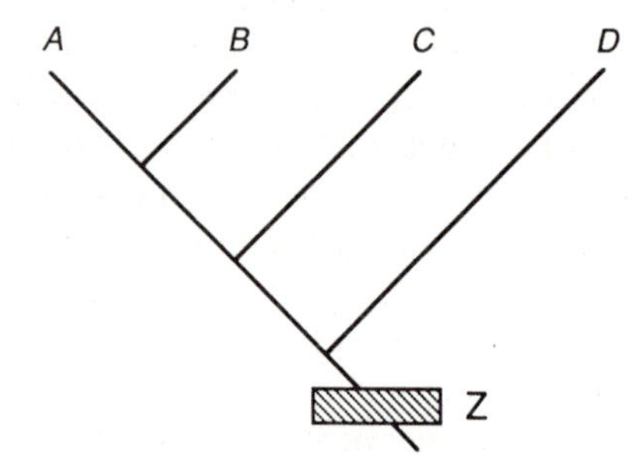

Fig. 2.1. Z marks a morphological change in an ancestral species. *A, B, C,* and *D* are terminal taxa that each inherit the new character state. Transformational studies examine changes in phenotype along the paths linking Z with the different terminal taxa. Relational studies examine the diversity of terminal taxa in the monophyletic group containing Z. (After Lauder 1981).

It is even possible to show the same transformations and relations occurring in a single monophyletic group. For example, the various routes to upper jaw protrusion in cichlid fishes discussed above may have been followed independently by adaptive radiations in different African lakes. Lauder's discussion offered a different and potentially useful phylogenetic perspective to comparative analyses for, as he presented it, Lauder considered only intrinsic features. That is, he was not concerned with how phenotypes evolve in response to external selective pressures (after all, several different types of upper jaw protrusion in cichlid fishes may be adaptively equivalent). Comparative biologists often relate phenotypes to extrinsic factors in the environment, but Lauder asked whether the origin of particular phenotypic novelties might inevitably result in the evolution of other predictable phenotypic changes. Environmental change may have selected for the particular feature in the first place but, once that feature evolved, it may have influenced strongly the course taken by subsequent evolution.

Lauder's approach does not mean that we should exclude extrinsic factors from historical analyses (see Liem and Wake 1985). Similar historical patterns may be detected following particular key innovations, but those changes may only be properly understood with reference to selective factors operating in the external environment (i.e. extrinsic factors). Take the example of the repeated adaptive differentiation of leaves into tendrils, hooks and spines (Hutchinson 1969). We could analyse separate phylogenies structurally in terms of transformational and relational patterns, but a biologically comprehensive understanding would include a consideration of the selective forces, such as predation and the need for climbing plants to reach the top of the canopy, which were operating when those structures evolved. Similarly with cichlid fish radiations in African lakes: presumably the particular feeding behaviours and associated morphologies that evolve in any radiation will have been in part determined by available food sources.

Irreversible evolution

Finally, some evolutionary routes may preclude return with the result that, when ancestral environments reappear, ancestral adaptations do not re-evolve (or at least do not re-evolve with facility). Groups of related species may thereby end up in the evolutionary equivalent of black holes from which there is no easy return. This topic of irreversible evolution is reviewed by Bull and Charnov (1985) and by Harvey and Partridge (1987). Godfray (1987) provides an instructive, if gruesome, example. Species of the Braconid hymenopteran genus *Apanteles* lay their eggs in lepidopteran caterpillars. More than half of 276 species for which Le Masurier (1987) could find data are solitary, with a single young feeding in its host. Many of the solitary species have long piercing mandibles which are used to kill fellow parasitoids in the same host. The other species are gregarious with typically 12–26 young per host, but occasionally up to 1200. Why, despite the availability of suitably sized hosts, do very few species of *Apanteles* lay broods of 2–11 eggs? Godfray's population genetic models demonstrate that genes for fighting should spread, and thus murderous mandibles should evolve, whenever competition for resources in a host gets intense. Once such mandibles have evolved, mothers are selected to lay just a single egg in a host because only one young can survive however many eggs are laid. If the environment changes and hosts grow larger over evolutionary time, the production of solitary aggressive young may remain the evolutionarily stable strategy. Indeed, in the simplest models considered by Godfray, the condition for genes for tolerance invading a population of fighters involves an Allee effect (individuals must have a higher fitness as one of a group than when alone). If Godfray's explanation for the dearth of parasite species laying few eggs is correct, we might expect: (1) a correlation between host and parasite biomass in gregarious species; (2) only solitary parasites to occupy small hosts; and (3) both solitary and gregarious species to occupy larger hosts. All three predictions hold (Godfray 1987). Furthermore, when large-bodied hosts are compared, the biomass of gregarious species per host is more than ten times that of solitary species (Le Masurier 1987).

Other examples of evolutionary routes along which it is easier to travel one way than the other include the transition between inbreeding and outbreeding. Once individuals in an inbred population begin to outbreed, recessive deleterious mutations will accumulate and inbreeding depression can become a potent force preventing the subsequent evolution of inbreeding. Currently, we lack a unifying framework that might help predict where we might find other cases of irreversible evolution (see Bull and Charnov 1985).

2.4 Afterwords

Although the comparative approaches that we shall be describing in Chapters 4 to 6 of this book generally seek evidence for similar phenotypic responses to similar selective forces, it should now be clear why that is not always what happens (see also Bock 1977, 1980). Most biological laws are limited in their generality and exceptions abound but, as the many examples cited in this book suggest, we believe that Bock (1980, p. 225) is throwing the baby out with the bathwater when he writes 'that the existence of these exceptions means that the comparison method of judging adaptations is not lawlike and hence invalid'! We have seen how different adaptive responses and time lags can result in imperfect correlations between character states and environments, but appropriate choice of taxa for analysis will usually help to unravel the factors involved. Phylogenetic niche conservatism, on the other hand, is more likely to lead to statistical problems: speciose taxa can bias statistical measures, such as correlation and regression, and degrees of freedom will be overestimated when species are used as the units for analysis.

If comparative studies are to explain the diversity among species, the roles of phylogenetic conservatism, phylogenetic time lags, and the diversity of adaptive responses must all be assessed. A rough measure of the extent to which evolutionary history has moulded a character is provided by comparing very closely related species living in different environments. If the variation among species in a character is high so that closely related species have evolved different phenotypes, evolutionary history has probably been of little importance in preventing change.

2.5 Summary

The careful use of phylogenetic relationships can help distinguish cause from effect and control for the influence of confounding variables in comparative studies. Closely related species tend to be phenotypically similar to each other as a consequence of at least three different biological processes: phylogenetic niche conservatism, phylogenetic time lags, and similar adaptive responses. Statistical degrees of freedom are easily inflated if niche conservatism and time lags are not taken into account when analysing comparative data, whereas different adaptive responses among phylogenetic lineages are best recognized by an appropriate choice of taxa for comparison.

3

Reconstructing phylogenetic trees and ancestral character states

What might have been is an abstraction
Remaining a perpetual possibility
Only in a world of speculation. (T.S. Eliot, *Burnt Norton*)

3.1 Introduction

We described in the previous chapter how living organisms come to contain information about their evolutionary history, and why this means that comparative analyses must utilize phylogenetic information. This chapter asks what phylogenetic information we need for comparative tests, and how we might obtain it. It will become evident that although defining the information that we need is fairly straightforward, getting it is another matter. In fact the overriding message of this chapter will be that in the absence of a good fossil record we can never be sure of evolutionary history. We often produce quite different pictures of the past by basing our reconstructions on evolutionary models that make different assumptions about the roles of processes such as natural selection and genetic drift. An important task for biologists is to define which models are based on the most biologically realistic assumptions.

A phylogeny, which we treat as synonymous with a phylogenetic tree, is a genealogical history of a group, hypothesizing ancestor-descendant relationships (Levinton 1988, p. 49). Comparative tests seek evidence for correlated evolutionary change between the states of two or more characters. As a consequence, comparative biologists need to know in what lineages and at what times evolutionary changes occurred. This means that we need to know the branching pattern of the phylogenetic tree which links the species in our sample, and the positions on that tree of specified evolutionary events. More often that not, the timings of branching patterns and evolutionary events are given in relative rather than absolute terms.

The structure of the phylogenetic tree is used to tell us when any pair of species last had a common ancestor. Usually, the phylogeny simply tells us which pairs of species had more recent common ancestors than other pairs. As data and techniques improve, however, inferred phylogenies increasingly include approximate dates of particular branching events so that the absolute rather than relative timing of occurrence of the most recent common ancestor is specified. The positions in the tree or timings of character state changes are usually provided by specifying ancestral character states at each node of the tree. If consecutive nodes have the same character state, it is assumed that no character change occurred along the branch linking the nodes, but if consecutive nodes have different character states we conclude that evolutionary change occurred along the branch. Our aim in this chapter, then, is to describe and assess techniques that are used to reconstruct both phylogenetic trees and ancestral character states.

The reconstruction of phylogenies has been the subject of considerable and often intemperate debate for many years (Felsenstein 1986; Hull 1988) and, more recently, the accumulation of molecular data has added a new level of interest. It is not our intention to review this debate, as such reviews have been undertaken regularly in the past by a variety of authors using different perspectives. Recent examples include Ridley (1986*a*), Hull (1988), and Sober (1989). Instead, we shall focus on the central issues which must be resolved if phylogenies are to be reconstructed with reasonable accuracy, and which bear upon determining hypothetical ancestral character states.

The next section, which considers the reconstruction of phylogenetic trees, asks which procedures for classifying organisms are most likely to result in the production of taxonomies that best describe phylogenetic relationships. Our conclusion is that cladistically defined taxonomies are usually the most suitable, but that some of the procedures commonly used by cladists can be improved upon to provide better estimates of true phylogenies. We go on to examine issues in tree reconstruction that are relevant to comparative studies. In particular, we show how the revolution in molecular biology can provide improved phylogenies, so long as the molecular information used is of the right kind and is from the right taxa. Finally, we discuss methods that can be used to reconstruct ancestral character states at the nodes of trees of known structure.

3.2 Reconstruction of phylogenetic trees

Once the need to use phylogeny in comparative studies has been accepted, the natural procedure has been to refer to standard taxonomies. Unfortunately, it is usually the case that traditional taxonomic relationships are not even *meant* to describe hypothetical phylogenetic relationships. For example, for reasons we shall describe below, common taxonomic practice classifies crocodiles and lizards together in a taxonomic group from which birds are excluded, even though crocodiles have a more recent ancestor in common with birds than they have with lizards. This is a crucially important point for the comparative biologist, because the use of taxonomies that are known to contradict phylogenetic relationships will often lead to incorrect conclusions from comparative analyses: garbage in, garbage out.

Which taxonomies best represent phylogenetic relationships and which are best avoided? Different schools of taxonomy are easily identified: there are pheneticists, cladists, transformed cladists, evolutionary taxonomists, and others. Most contemporary taxonomies were constructed by evolutionary taxonomists and, as we shall see, their taxonomies explicitly were not designed to describe phylogenies. For entertainingly pithy accounts of the differences among the various schools, see Ridley (1983*b*, 1986*a*).

3.2.1 The schools of taxonomy

Evolutionary taxonomists have been careful to distinguish among the reasons for phenotypic similarity among species. Similarity can result from either convergent evolution, parallel evolution, or identity by descent (see Fig. 1.5). For evolutionary taxonomists, only those characters that are identical by descent should be used to decide upon taxonomic affinity. Convergently evolved characters, that is characters that are not homologous, should not be used to place species in the same taxonomic group. So much for similarity among characters, but what of phenotypic differences among species? Here, evolutionary taxonomists take a different stand: phenotypic difference often takes precedence over phylogenetic relatedness in the production of a taxonomy. One consequence of using phenotypic divergence to construct classifications is that some taxonomic groups are not monophyletic, by which we mean that they do not contain all the descendants of a particular common ancestor.[5] The birds, lizards, and crocodiles mentioned above are a case in point. Birds and crocodiles are more closely related phylogenetically, than either group is to lizards. However, because birds with their wings, feathers and beaks have evolved to look quite different from crocodiles, birds are placed in the Class Aves while crocodiles are placed with the phenotypically similar lizards in the Class Reptilia.

If there is one thing that virtually all comparative biologists are agreed upon, it is that taxonomic groups should be monophyletic, because only

---

[5] We define a monophyletic group as one containing all the descendants of a particular common ancestor, thus following Hennig's (1966) definition rather than those of Simpson (1961) and Mayr (1969). See Holmes (1980) and Ridley (1986*a*) for useful historical accounts of the concept.

then will a hierarchical taxonomy be isomorphic with a phylogenetic tree. Nevertheless, comparative biologists often use standard taxonomies derived by evolutionary taxonomists as substitutes for phylogenetic relationships. Yet, for example, 'only about half of the classes of the Chordata are thought to be monophyletic . . . and even within the Mammalia orders such as Insectivora and Carnivora are believed not to be monophyletic' (Felsenstein 1985*a*, p. 7). Many comparative studies of mammals continue to use Simpson's (1945) 'evolutionary taxonomy' as their baseline, despite the availability of classifications of the mammalian radiations which more accurately represents phylogeny (e.g. Eisenberg 1981, and examples in Benton 1988*a*, *b*). The classifications of evolutionary taxonomists are not satisfactory for comparative studies.

Pheneticists do not even consider phylogeny as a factor in the construction of their taxonomies. Instead, pheneticists construct taxonomies on the principle of phenotypic similarity. All characters for which information is available, whether evolutionarily convergent or divergent, are included to construct measures of phenotypic similarity. Then statistical methods for detecting clusters are used to produce a hierarchical classification (Sokal and Sneath 1963; Sneath and Sokal 1973). Different cluster statistics produce different taxonomic groupings, and none of them even purport to produce phylogenetic groupings. Furthermore, even if an appropriate cluster statistic could be defined, phenetic methods would still be inappropriate for identifying phylogenetic groupings. One important reason is that, as we shall discuss in the next section, character states that are primitive to a phylogenetic group should not be used for inferring relationships *within* that group (see Fig. 3.1). Pheneticists, however, treat

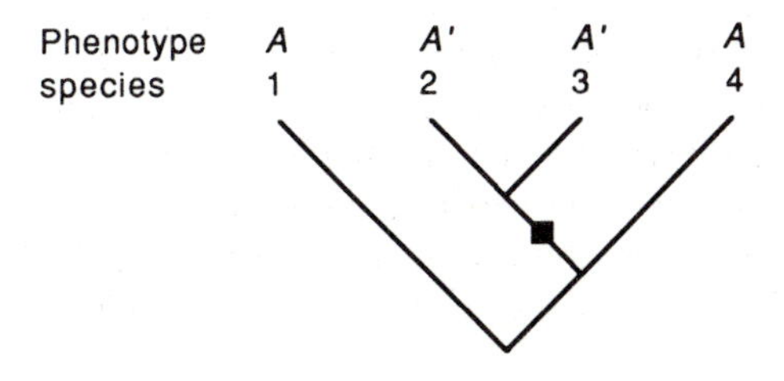

Fig. 3.1. A phylogenetic tree in which ■ represents an evolutionary transition of a character from state A to state A'. Phenetic similarity on the basis of that character would correctly group species 2 and 3, but incorrectly group species 1 with 4. An analysis using only shared derived character states would correctly group species 2 and 3, but leave other groupings undetermined.

derived and primitive character states equally.[6] The techniques of pheneticists, therefore, are not satisfactory for our purpose.

Cladists attempt to use phylogenetic branching as a basis for classification. Historically, cladists have used a set of rules that Willi Hennig (1966) devised with the express purpose of reconstructing phylogenetic trees as accurately as possible. Hennig's central claim was that a hierarchy of shared derived characters can be used to recognize a hierarchy of monophyletic groups which, itself, is a hierarchy of recency of common ancestry (see Fig. 3.1). *Pattern cladists* (Nelson 1979; Patterson 1980; Nelson and Platnick 1981, 1984) use Hennig's rules without the underlying philosophy of reconstructing phylogenetic trees, so they need not concern us here. Because phylogenetic trees are our goal, Hennig's approach makes a good start.

3.2.2 Shared derived characters as the basis for inferring phylogenetic relationship

Character states that are primitive to a monophyletic group are not useful for inferring evolutionary relationships within that group (see Fig. 3.1). Ridley (1986*a*, p. 54) gives a biological example: 'Suppose we wish to classify a baboon, a crocodile, and a cow relative to each other, and study the states of their limbs. The baboon and the crocodile have five toes, the cow two; but the fact that a baboon and a crocodile both possess a pentadactyl limb—the ancestral condition for tetrapods—is not evidence that these two species are phylogenetically closer to each other than is either to the cow. Shared ancestral characters do not reveal phylogenetic relationship'. Indeed, it is because lizards and crocodiles retain more ancestral character states in common than either group does with birds that evolutionary taxonomists classify lizards together with crocodiles in the Reptilia and birds separately in the Aves.

It was appreciated by some taxonomists early this century that only shared derived characters should be used to establish phylogenetic relationship (see Felsenstein 1982). For example, Mitchell (1901, pp. 181–2) described the practice as 'merely a codification of criteria in common employment among naturalists' and Le Gros Clark and Sonntag (1926, p. 478) declared that the aardvark *Orycteropus afer* 'shares primitive features with most Edentata, but these do not imply relationship'. However, Hennig's (1966) book, which prescribed a taxonomic procedure that was explicitly based on the shared derived character criterion, has been the most influential factor in the general acceptance of this procedure.

[6] It might be argued here that it is not possible to determine which character is derived and which primitive until a phylogeny is known! As we shall see in Section 3.2.5 and 3.3, a variety of methods are available to determine ancestral character states.

Hennig's procedure assumes that ancestral states for each character are known, that evolution of character states is irreversible, and that each character can change only once in a phylogenetic tree. Consider the data set given in Table 3.1 (simplified from Felsenstein 1982). All five characters (1 to 5) recorded in five species (*A* to *E*) can occur in either of two states, 0 or 1, of which 0 is ancestral.

Table 3.1 Each of five characters (numbered 1 to 5) which can occur in one of two states (0 or 1) are scored in each of five species (labelled A to E). A phylogenetic tree based on change in the first four characters but making certain assumptions (see text) about the evolution of character states is given in Fig. 3.2.

Species	Character				
	1	2	3	4	5
A	1	1	0	0	0
B	0	0	1	0	1
C	1	1	0	0	1
D	0	0	1	1	0
E	0	1	0	0	1

Given Hennig's assumptions, each character defines a monophyletic group. For example, character 1 defines (*AC*) and character 2 defines (*ACE*). Taken together, the first three characters define a unique phylogenetic tree, with which the fourth character also agrees (Fig. 3.2).

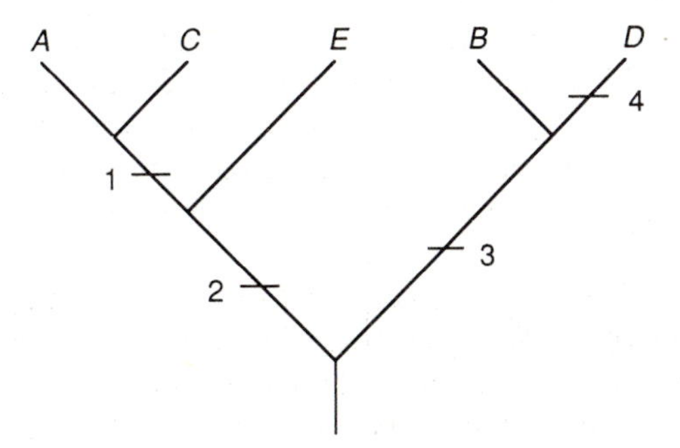

Fig. 3.2. The unique phylogenetic tree determined by the evolution of characters 1 to 4 in Table 3.1, under the assumption that each character can change state only once in the tree. The transition branch for each character is marked.

However, character 5 defines the monophyletic group (*BCE*) which contradicts the phylogenetic tree constructed by characters 1 to 4. There is no phylogenetic tree that can be constructed using all characters with Hennig's criteria. This is what Felsenstein (1982, p. 381) refers to as 'Hennig's dilemma'. How can it be resolved?

3.2.3 Resolving Hennig's dilemma: parsimony, compatibility, and maximum likelihood

Relaxing Hennig's assumptions

The resolution of Hennig's dilemma reduces to a matter of philosophy. If we use a hypothetico-deductive approach to tree-building, Hennig has put forward a series of assumptions which together constitute an hypothesis that the data can falsify. We might then change the hypothesis by relaxing one or more of Hennig's assumptions. We shall discuss a number of such possibilities below. For the purposes of the present discussion, we consider one scenario. What would be the consequence of relaxing Hennig's assumption that each character can evolve the derived state on only one occasion?

For the example discussed above (Table 3.1), if we allow characters to evolve their derived states on more than one occasion, we should be faced with a new problem: *all* potential trees can be assigned character states so that they fit the data! All trees could be chosen so that the state for each character at each node is 0. All evolutionary changes for characters 1 to 5 would then occur in the terminal branches linking each species to its most recent node. There are 11 origins of derived character states in such trees, corresponding to the 11 derived character states labelled 1 in Table 3.1. Some of the trees could be drawn with fewer state changes. For example, the tree in Fig. 3.2 would need three additional changes to deal with the evolution of 0 to 1 in character 5. Those changes would occur along the terminal branches leading to species *B*, *C*, and *E*, making seven character state changes in all. We need a criterion for choosing among all the different trees. One possibility is to choose the tree with the least number of character changes, summed across all characters. This would be a parsimony criterion.

Felsenstein (1982) reviews various relaxations of Hennig's assumptions, with their associated parsimony criteria. First, the same derived characters may be allowed to evolve more than once in the tree ('Camin-Sokal parsimony', after the method used by Camin and Sokal 1965); the most parsimonious tree is the one with the smallest number of separate derivations. Second, derived characters can originate only once in a phylogeny but may be lost many times ('Dollo parsimony', named loosely after Dollo's law which states that complex characters will not have evolved more than once); the most parsimonious tree is the one in which derived characters are lost on the smallest number of occasions. Third, both multiple derivations and multiple losses are allowed with the consequence that the ancestral character state for the common ancestor of all the species in the tree is not automatically specified; the most parsimonious tree is the one which shows the smallest number of character state transitions.[7] A fourth and final method allows an intermediate or polymorphic state (01) in any character; the most parsimonious tree is that with such polymorphisms persisting for the shortest amount of time. It is assumed that the polymorphism between any two character states can arise just once and that the derived character state can be reached only from the polymorphic condition (Inger 1967; Farris 1978; Felsenstein 1979). Other rules of character change are discussed by Gillespie (1986*a*) and by Maddison and Maddison (1989).

Hypothetico-deductive versus statistical methods for deciding among phylogenetic trees

The procedure just outlined is: (1) to decide on a set of rules (specified by Hennig's assumptions in the case discussed above); (2) if necessary, to relax one or more of the rules so that a number of trees are possible; (3) to choose that tree for which, summed across all characters, Hennig's original assumptions are broken the smallest number of times (a parsimony criterion). Alternatively, we might have decided to choose the tree containing the largest number of characters that break none of the rules (a compatibility criterion).

Indeed, the phylogenetic tree of choice could be decided upon using both parsimony and compatibility criteria. For example, a parsimony criterion, such as the minimum number of character state changes along the branches of the tree, might be used for choosing the ancestral state at each node for each character. A second criterion, such as that tree which is the most parsimonious for the greatest number of characters, could be used to decide upon the final phylogenetic tree. This would be a type of compatibility criterion. The most compatible tree would not necessarily have been chosen if the second criterion had been to select the most parsimonious tree, defined as that tree requiring the minimum total number of changes summed across all characters. For example, Table 3.2 gives the results of two alternative tree structures when a number of taxa were classified according to each of three characters. *A* is the chosen tree

---

[7] With multiple states, this model is either 'Wagner parsimony' (defined as the 'Wagner method' by Kluge and Farris 1969) when character states are in an ordered sequence (0, 1, 2, 3 . . .) with only single-step transitions allowed, or 'Fitch parsimony' when any state-to-state transition is possible.

when our compatibility criterion is used in the second stage of the process, but *B* is the chosen tree when parsimony is used. An important point to be made here is that, under our compatibility criterion, the number of characters for which the chosen tree is the most parsimonious is a subset of all the characters used, and the number of changes in the tree required by characters not belonging to that subset may be considerable.

Table 3.2 Several species are classified according to each of three characters. The minimum number of character state changes required to produce the same two hypothetical phylogenetic trees differs for each character. Summed across all characters, tree B is the most parsimonious tree, requiring a total of 13 character state changes. However, using a simple criterion of compatability, tree A which is the most parsimonious tree for both characters 1 and 2 would be chosen.

Character	1		2		3	
Tree	A	B	A	B	A	B
Character state changes	3	4	3	4	8	5

However, there are problems associated with applying either compatibility or parsimony within a hypothetico-deductive framework because the hypothesis of choice is falsified by the chosen tree if any rule specifying that hypothesis is ever broken. This can lead to what may seem to be an awkward situation: 'that phylogenetic hypothesis which has been rejected the least number of times is to be preferred over its alternates' (Wiley 1975, p. 243). Nevertheless, as we shall see below, parsimony and compatibility methods do, in fact, have an important part to play in phylogenetic tree reconstruction and in the reconstruction of ancestral character states. The important issue for our discussion will be to find the conditions under which they can be expected to yield the true phylogeny.

A different approach (Edwards and Cavalli-Sforza 1964, and championed in particular by Felsenstein 1973*a*, *b*, 1985*b*, *c*), is to view phylogenetic tree reconstruction as a statistical problem. One statistical procedure is to seek the maximum likelihood tree, which is that phylogenetic tree with the highest probability of having produced the observed data under a certain set of probabilities of character change. This statistical approach fits nicely with biological intuition. After all, provided that all transition probabilities from one state to another are greater than zero, no tree is impossible, it's just that some trees are more likely than others. The problem is put into perspective by considering gene frequency data. Felsenstein (1985*b*, pp. 300–301) points out that, given appropriate levels of mutation, selection and drift: 'Any given pattern of gene frequencies could arise on any given phylogeny, although with lower probabilities on some phylogenies than others. The notion of falsification is called into question', because no tree could be falsified by any pattern of gene frequencies.

Maximum likelihood is a statistical estimation procedure which, in the case of phylogeny reconstruction, must be based on a model of evolutionary change. Maximum likelihood estimators have the property of consistency over a wide range of evolutionary rates and tree topologies. This property ensures that the estimator is increasingly likely to yield the correct tree as more data are collected. However, maximum likelihood solutions can be technically and computationally difficult to achieve even under some of the simplest (and most unrealistic) models of change. For example, Edwards and Cavalli-Sforza (1964) developed their method for constructing evolutionary trees based on the parsimony criterion of minimum evolution because they were unable to produce a suitable maximum likelihood solution for evolution under genetic drift. As a result, it has become common practice to use compatibility or parsimony procedures because they are tractable even though evolution does not necessarily proceed that way. If we are to use compatibility and parsimony procedures instead of maximum likelihood, we need to know the circumstances under which they yield maximum likelihood solutions (or are acceptable under some other statistical criterion), and then determine if those circumstances are biologically realistic.

When compatibility and parsimony procedures converge on the correct tree

Compatibility finds that tree which is compatible with the greatest number of characters given the rules of character change used (Estabrook 1972, 1980; Le Quesne 1969). The tree in Fig. 3.2, for example, is compatible with 4 of 5 characters under Hennig's assumptions for the ways characters can change. Character 5 would have needed to change three times to be compatible with the tree, but Hennig's rules say that it cannot have done this. When many characters are used in the construction of a tree, very few characters may be compatible with the chosen tree so that, in fact, the majority of characters do not help define the tree structure. Methods to incorporate information from non-compatible characters were pioneered by Estabrook *et al.* (1977) and by Estabrook and Anderson (1978). It might seem that compatibility methods return us to a hypothetico-deductive approach, but Felsenstein (1979, 1981*a*) has discussed some conditions under which compatibility methods provide maximum likelihood trees.

The latter are known in these circumstances to converge on the true trees as more data are collected. One circumstance involves the case where some characters change sufficiently rarely to produce only a single origin of a derived state while others produce many origins of the derived states. The characters with multiple origins contain little useful information and can, indeed, mislead taxonomists; they may be excluded by compatibility methods. A useful discussion of the role of compatibility methods in the reconstruction of phylogenies, including references to examples from buttercups to birds, is given by Meacham and Estabrook (1985).

Another possible resolution of Hennig's dilemma involves using parsimony which, in the case discussed in Section 3.2.2, means relaxing one of two of Hennig's assumptions: either a character state can arise more than once or character change is reversible. The most parsimonious tree is that which minimizes the number of additional evolutionary changes in the tree. Interestingly, Hennig (1950, 1965, 1966) prescribed neither compatibility nor parsimony. Instead, he recommended re-checking the character states in the original data. The hypothesis should never be refuted!

But is the most parsimonious tree most likely to be the correct tree? Not always (Cavender 1978; Felsenstein 1978; Hendy and Penny 1989). As an example, Felsenstein (1978, 1983) produced a phylogenetic tree containing four living species connected by branches along which rates of character state change had been different. The tree is reproduced in Fig. 3.3 and, for the sake of discussion, is assumed to be the correct tree. The probability of

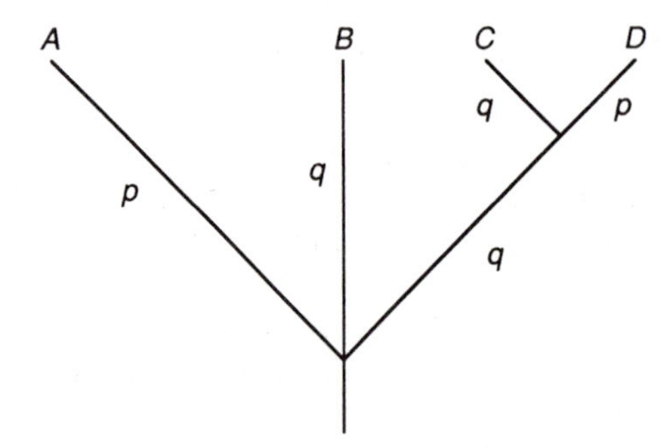

Fig. 3.3. A phylogenetic tree that is likely to produce data that will cause parsimony or compatibility methods to give misleading results. The correct tree is shown, with p and q being probabilities of character state change along the different branches of the tree. The wrong tree is likely to be determined when p is much larger than q because then B and C are likely to be more similar than are C and D. Details are given in the text. (After Felsenstein 1978, 1983).

character state change in any single branch is either p or q, but p is much larger than q. Transitions from character state 0 to state 1 are as likely as transitions from 1 to 0. When Wagner parsimony is used (forward and backward transitions allowed) on a single dichotomous character to reconstruct the tree from the character states of the extant species, the reconstructed tree is more likely to be incorrect than correct.

This is because a derived character found in only one species could have arisen in any terminal branch, thus providing no information concerning relationships among species. However, of the informative combinations of character states, the most likely for *ABCD* are 0110 or 1001 simply because *B* and *C* are likely to share the same character state if q is low enough: *B* would be classified with *C*, or *A* with *D*. Felsenstein demonstrates that if a tree was reconstructed using Wagner parsimony, so long as $p^2 > q(1\text{-}q)$, the most parsimonious tree probably would not show *C* being classified with *D*. Indeed, as more characters are included, each with the same rates of change, it becomes less and less likely that the most parsimonious or the most compatible tree will be the correct tree (Felsenstein 1979).

Most comparative studies have been analysed using statistical methods that assume equal branch lengths and equal rates of transition along each branch (as we shall see in Chapters 4 and 5). The dangers of making these assumptions without supporting evidence are clear from Felsenstein's example. In practice, rates of change for many characters may both covary with each other and vary among lineages for a variety of reasons. For example, neoteny or other changes in developmental timing may result in the seemingly co-ordinated evolution of many characters. Similarly, changes in rates of hormone production can affect the development of many characters. And, as we saw in the previous chapter, key innovations (Lauder 1981; Liem and Wake 1985) may herald predictable sequences of change in many characters.

Felsenstein (1983, p. 324) concluded that 'parsimony methods are well justified if the rates of change are sufficiently small or if they are sufficiently equal in different lineages'. In the example given above, rates of change differ among lineages. Sober (1989, p. 198), argues that Felsenstein's two conditions may be over-restrictive: because such conditions may be sufficient to make parsimony produce a maximum likelihood tree, this does not mean that parsimony methods would fail to produce approximations to maximum likelihood trees under other modes of evolution with higher rates of character change. Sober's distinction is between sufficient and necessary conditions. Hendy and Penny (1989), however, report that parsimony can converge on the wrong tree even when rates of change are equal, or when rates are unequal but 'low', for trees with more than four species. Hendy and Penny report that this is most likely to occur when the root of the tree divides into a single species on one side, with the remaining species on the other side (as in Fig. 3.1). Parsimony may converge to the correct tree if one or more outgroups to the single species can be added to the tree.

In summary, we can never be sure that we have constructed the correct tree, and so we must rely on methods that will at least converge on the correct tree as more information is collected. This is the property of consistency. Maximum likelihood techniques for inferring trees are well justified in recognizing that any phylogenetic tree is possible (given that the probability of character change is not zero), and give consistent estimates under a very wide range of conditions. Unfortunately, maximum likelihood can quickly become computationally impractical for large numbers of species. Parsimony and compatibility methods can often be counted on to give consistent estimates of the true phylogeny when the probability of character change is small and roughly equal throughout all lineages (but there problems even here, as pointed out by Hendy and Penny 1989). Where there is evidence for parallel evolution or a character shows a large number of independent transitions, then these assumptions are questionable, and the resulting phylogeny should be treated with caution. Characters to be used for tree reconstruction are best chosen bearing those principles in mind. As we shall see in the next section, appropriate choice of DNA or amino acid sequence data may fulfil the twin criteria of low and equal rates of change.

3.2.4 Molecular data

Different DNA sequences diverge at different rates

A frequent assumption made when analysing data from macromolecules to produce phylogenetic trees is the constancy of the molecular clock: divergence at the molecular level occurs at a reasonably constant rate (in terms of absolute time or number of generations). A critical evaluation of the clock concept for the evolution of a variety of macromolecules is provided by Thorpe (1982). Fortunately, the clock runs at different rates for different types of macromolecule and for different regions of the same molecule, so an appropriate molecule may be picked for the question being asked.

When using comparisons among sequence data for reconstructing phylogenetic trees, sequences should be chosen which have diverged sufficiently so that the taxa in the sample can be discriminated from each other, while still bearing varying degrees of resemblance through common ancestry. Ribosomal RNA sequence analysis is useful for discriminating distant levels of relationship because of its relatively conservative structure, and has been used to throw new light on the origin of molluscs (Ghiselin 1988). Solution hybridization of total single-copy nuclear DNA

(DNA hybridization) provides useful data for closer relatives; it cannot be used to provide evidence about relationships between different phyla but, among birds, it works at all taxonomic levels, at least above the tribe (Sibley and Ahlquist 1983). Some regions of the mitochondrial DNA genome can be used to distinguish even closer phylogenetic relatives, particularly among species and populations (Wilson *et al.* 1985). Finally, the use of appropriate probes to either detect DNA single-locus polymorphisms (Quinn and White 1987) or distinguish between hypervariable DNA sequences, such as DNA fingerprinting (Jeffreys *et al.* 1985; Jeffreys 1987), are rapidly replacing analyses of electrophoretically detected enzyme polymorphisms to determine familial relationships within local single-species populations (see Burke 1989).

One particularly comprehensive application to date of molecular techniques for phylogenetic inference is Sibley and Ahlquist and Monroe's (Sibley and Ahlquist 1983, 1984, 1985; Sibley, Ahlquist and Monroe 1988) studies of birds and hominoid primates using nuclear DNA hybridization. Radioactively labelled DNA strands of about 500 nucleotides in length from a focal species were hybridized with similar strands from a number of phylogenetic relatives. The rate of dissociation of the duplexes provided a measure of similarity between the sequences. These molecular phylogenies agree, for the most part, with those derived from morphological and biogeographical data (Diamond 1983), but there have been some surprises. For example, Australian passerines seem to be more closely related to each other than they are to passerines from other continents, thus suggesting that previous classifications based on morphological comparisons had failed to detect considerable convergent morphological evolution. While DNA hybridization studies offer considerable promise for reconstructing phylogenies by estimating the amount of divergence among single-copy fragments of vertebrate genomes, we should be wary about treating many published studies as providing more than suggestive evidence (see Sarich *et al.* 1989). In particular, some measures of divergence that have been used may not be reliable and it is not possible to calculate alternative measures in the absence of published raw data.

For several reasons, mitochondrial DNA is proving a particularly powerful tool in phylogenetic tree reconstruction (Avise *et al.* 1987). For example, it is easy to isolate and assay, it has a simple genetic structure (lacking repetitive DNA, transposable elements, pseudogenes, and introns), and it is inherited without recombination. The order of genes (rather than the sequence of bases within them) is stable among three mammalian genera and an amphibian but differs between them and *Drosophila*, thus providing a possible route for phylogenetic reconstruction among very distantly related taxa (Harrison 1989). Nucleotide sequences can evolve about ten times faster than in most single-copy nuclear DNA (Vawter and Brown 1986), although this relative rate varies among taxa, in large part because of varying rates of nuclear DNA evolution (Britten 1986; Moritz *et al.* 1987). Avise *et al.* (1987) describe mitochondrial DNA phylogenetic trees as 'self pruning' because mitochondria are usually maternally inherited (for exceptions see Kondo *et al.* 1990). For example, if females produce an average of one and a variance of five daughters according to a binomial distribution, then all mitochondria are likely to be inherited from the same foundress after $2n$ generations, where n is the number of females in the population (Avise *et al.* 1984).

Mitochondrial DNA has already established its credentials for phyloenetic inference among lower level taxa, for it has succeeded in several instances where other techniques have failed (see Moritz *et al.* 1987). In the past, restriction site maps or restriction fragment mobilities have been used for most studies, but in the future we can expect base sequence data to become more widely available. The reason for this optimism is the new-found ability to produce multiple copies of particular DNA fragments using the polymerase chain reaction. (In the past it was necessary to 'clone' the DNA using recombinant DNA technology, and to propagate it using *Escherichia coli*. With PCR, these steps are bypassed.) For example, using the method, Golenberg *et al.* (1990) have sequenced an 820 base pair DNA fragment from a 17 to 20 million-year-old magnolia (*Magnolia latahensis*) chloroplast gene. When they compared the sequence cladistically with sequences from homologous regions in other species, they found that it never grouped outside those of the other Magnoliidae that were examined. The polymerase chain reaction was introduced in 1985 (Saiki *et al.* 1985) but has now been both simplified and automated to the extent that large quantities of DNA can be obtained from a single molecule (Li *et al.* 1988), and sequenced directly (McMahon *et al.* 1987).

As lineages become separated for longer times, repeated substitutions at the same sites result in lower rates of sequence divergence (see Fig. 3.4 after Hixson and Brown 1986; Brown *et al.* 1982). Under such circumstances, a shift of attention to analysing base substitutions only from areas of the mitochondrial genome that change less rapidly (either RNA-coding sequences or second codon positions of protein genes) can allow reasonable resolution at higher taxonomic levels (Moritz *et al.* 1987). However, there is increasing evidence that rates of mitochondrial DNA divergence differ among lineages (e.g. Templeton 1987; Moritz *et al.* 1987). We have already seen how differing rates of evolution pose problems for phylogeny reconstruction based on parsimony analyses (see above and Fig. 3.3).

Comparisons between the structure of protein molecules have also been used for phylogenetic inference, sometimes as direct amino acid sequence data (e.g. Fitch and Margoliash 1970) and sometimes more indirectly as

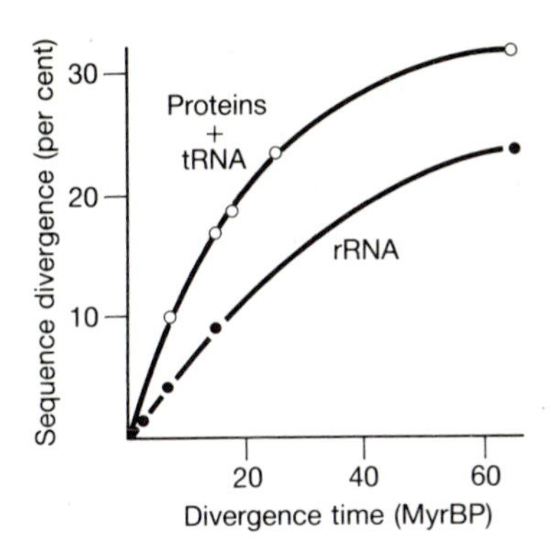

Fig. 3.4. The dynamics of mitochondrial DNA sequence divergence for primates, cow, and mouse (after Moritz *et al.* 1987). As lineages become separated for longer periods of time, repeated substitutions at the same site causes the apparent substitution rate to drop dramatically. Different parts of the mitochondrial DNA genome change at different rates, with regions coding for ribosomal RNA (rRNA) changing more slowly than those coding for transfer RNA (tRNA) and proteins.

immunological distances (e.g. Sarich 1977; Wilson *et al.* 1977; Wyles and Gorman 1980).

Models of molecular evolution

While, as we have seen, the molecular revolution is beginning to supply us with new data that will allow the increased resolution of phylogenetic trees, this does not mean that the data will analyse themselves. Indeed, there are formidable statistical problems associated with using those data correctly, and the development of better models based on an understanding of the forces involved in molecular evolution has become imperative (Gillespie 1986*a*, *b*, 1987; Friday 1989).

Gillespie (1986*a*) pointed out that many models for molecular evolution, whether constructed to deal with nucleotide, codon, or amino acid substitutions, assume that substitutions at different sites are independent events (in fact, independent stationary Markov processes). If that assumption is correct, queuing theory predicts that substitutions along lineages of set lengths will conform closely to a Poisson process. The ratio of the variance in the number of substitutions to the mean (R) should be one under a Poisson process. However, for protein sequence data at both nuclear and mitochondrial loci, R is usually much greater than one, with a modal value between 2.5 and 3.5. In a study of DNA sequences, the range is appreciably greater, extending to 35.0 for amino acid substitutions and to 19.0 for silent substitutions (Gillespie 1986*b*). This means that the variance in the rate of evolution is at least an order of magnitude greater than expected from the neutral theory for some sequences (Gillespie 1987). Gillespie (1986*a*) points to four possible explanations for the deviations from a Poisson process. First, evolution within lineages could occur in bursts. Second, rates of evolutionary change could vary among lineages. Third, single mutations might affect more than one substitutional site. Fourth, correlated or compensatory substitutions occur. Distinguishing among these explanations and the suggestion that Gillespie's 'set lengths' were incorrect (Esteal 1990) is an important task.

Many interpretations of molecular evolution, as well as our understanding of the timing of branching points, depend on the notion of a molecular clock. As Gillespie (1986*a*, p. 659) points out, the evidence that R is greater than one 'does not argue against the molecular clock. Rather, it argues against using a Poisson process as a model of the clock'. Of the four explanations given above for the deviation of molecular evolution from a Poisson process, only the one that suggests rates of evolutionary change differing among lineages is at odds with the usual working definition of the molecular clock (see Thorpe 1982).

Many of the patterns already described, as well as others, argue strongly against purely neutral theories for molecular evolution. One alternative among the many possible theories which might account for some of the patterns in the data by allowing a role for selection was developed by Gillespie (see Gillespie 1987). The central concept is that of a 'molecular landscape'. At any point in time, the fixed allele at a locus is the most fit in the current environment among those alleles one mutational step away. When the environment changes, a previously deleterious allele may become favoured, and become fixed in the population. Now, a new set of mutants can be produced which were two mutational steps away from the previously fixed allele. One of these new mutants may be fitter than the currently fixed allele, and become fixed in the population. The process will continue until all the mutants that are one mutational step away from the currently fixed allele are less fit than it is. It can be seen that, following a change in the environment, this simple model can produce bursts of evolution, thus explaining deviations from the Poisson process outlined previously.

Almost all models of molecular evolution to date have been based on the neutral allele theory. If we are to meet the challenge of specifying models of molecular evolution, the demonstrated inadequacies of the strict neutral theory should prompt the development of realistic alternatives. Despite the 'extraordinary mathematical difficulties that these models present' (Gillespie 1987, p. 34), we hope that this section has demonstrated the critical role that they are destined to play in the phylogenetic and comparative analyses of the future. Phylogenetic trees and ancestral character states must be reconstructed on some theoretical foundation.

DNA–DNA hybridization data and the principles of cladism

The sequence of base pairs along the genome can be treated like any other character. If we knew the sequence along homologously derived sections of the genome for different species, we could use the agreement or disagreement at each of the sites as a character for cladistic purposes, according to Hennig's shared-derived criterion (e.g. Wölters and Erdmann 1986; Ghiselin 1988).

However, one technique for assessing the similarity of genomes from different species, DNA–DNA hybridization, has been criticized on the grounds that it cannot distinguish primitive from derived characters. Thus, the strength of the bond between between two strands of DNA will be a function of the number of sites at which they agree, regardless of whether the similarity is primitive or derived. The result, then, critics assert, is that DNA–DNA hybridization data have all the problems of phenetic approaches to classification. Not surprisingly, this charge has been denied by Sibley and Ahlquist (1987).

The debate over whether DNA–DNA hybridization identifies shared-derived characters revolves around the idea of how well the technique assesses total evolutionary change, which Springer and Krajewski (1989), whose arguments we summarize here, denote T. Assume that all genomes to be compared are equivalent in the sense that all sites in the genome of one species have homologous sites arranged in the same sequence in the genome of the other species. We wish to measure the pairwise differences between nucleotide sequences of a pair of species. The measure T is defined as a tally of all the point mutations that have occurred since the pair of species last had a common ancestor. T would include convergence events and is, by definition, a measure of total evolutionary change between two species. T is also an inverse measure of shared-derived characters: higher values of T, given our assumptions about genomic equivalence, mean fewer shared-derived characters.

Now, define a measure D, which is a simple tally of the number of homologous sites which differ in sequence between the two species. D will not count convergent events, and so will underestimate T. D, then, is a non-additive measure of the distance between two species. DNA–DNA hybridization can be thought of as an analogue measure of D. Thus, if we know the function linking D to melting points, we can convert temperature to genetic distance. This is what Sibley and Ahlquist attempt to do.

Viewed in this light, the debate about hybridization measures comes down to whether or not we believe that convergent evolution of nucleotide sequences is a common thing. If it is, and if it differs among lineages, hybridization measures will be distorted. If, however, convergence of nucleotide sequences is not common, hybridization measures will provide a good indication of the number of shared-derived characters between two lineages, even though the measure does not count individual characters. Springer and Krajewski (1989, p. 314) conclude that if hybridization measures are additive the 'essence of cladistic intent (i.e. that net amounts of synapomorphy [shared-derived characters] are evidence of phylogenetic relationships) is not compromised, even though precise identification of character states is lacking'.

Cases of both parallel and convergent evolution at the molecular level have, in fact, been identified (e.g. Romero-Herrera *et al.* 1978; Liao *et al.* 1986; Stewart *et al.* 1987). In their study of myoglobin amino acid sequence differences among mammals, Romero-Herrera *et al.* (1978, p. 63) discussed two reasons for non-divergent evolutionary change. First, only a particular subset of all possible amino acid substitutions is consistent with myoglobin retaining its function as both a short- and long-term oxygen store. Second, not all of those substitutions can be accomplished by a single base change. However, once one base change has occurred, this may open new avenues for subsequent change. As a consequence 'constraints demanded by the functional morphology of the molecule itself and the constraints of the genetic code . . . contribute to parallel change in different lineages'. Furthermore, Romero-Herrera *et al.* found evidence to suggest that particular amino acid sequence changes in the myoglobin molecule may be favoured by a species' lifestyle. For example, the aquatic mammals (Cetaceans and Pinnipeds) seemed independently to have evolved an excess of arginine residues, any adaptive significance of which remains unknown.

Another example of apparent evolutionary convergence at the molecular level comes from Stewart *et al.*'s (1987) comparative study of lysozyme sequence data. Lysozyme is used to fight invading bacteria and many mammals have moderate to high levels in tears, saliva, white blood cells, and tissue macrophages. However, foregut fermenters such as langurs (*Presbytis entellus*) and cows also use lysozyme to digest bacteria that pass from the fermentative foregut into the true stomach. The *c* class lysozyme of foregut fermenters must work at low *p*H and be unusually resistant to breakdown by pepsin. The true phylogenetic tree, which puts cows closer to horses and langurs to baboons, is not the most parsimonious tree accounting for amino acid sequence divergence in *c* class lysozyme among the four species. Instead, langurs are more similar to cows than they are to baboons. Two alternatives to evolutionary convergence—horizontal transfer of genetic material between the ancestor of a cow and that of a langur, and gene duplication with homologous copies of the lysozyme coding gene being activated in the langur and the cow—were effectively discounted by Stewart *et al.* (1987).

Even though there is no simple way to estimate the degree of convergence at the molecular level using DNA–DNA hybridization, there is little doubt that most sequence evolution is predominantly divergent (Stewart *et al. 1987*). Sibley and Ahlquist (1987) do not believe that convergence is a serious problem. However, Felsenstein (1984) suggests that additivity of hybridization distances must be supported empirically, while Wilson *et al*. (1977) discuss some early evidence for additivity in hybridization data.

3.2.5 Rooting the tree

Our discussion up to this point has assumed that we know the origin or root of the phylogenetic tree. However, if characters can change state reversibly during evolution, as they do under the assumptions of Wagner parsimony, roots of phylogenetic trees are not automatically specified as they are under Hennig's original scheme, where the ancestor is a species with all characters in the ancestral state. Unrooted trees obtained by Wagner parsimony were termed 'Wagner Networks' by Farris (1970). When methods are available to identify ancestral forms and thereby root the tree, 'Wagner Networks' become 'Wagner Trees'. Unrooted trees may be adequate for many non-directional comparative studies, but for directional comparisons we must be able to distinguish ancestral from derived nodes (for a discussion of the difference between directional and non-directional comparisons, see Section 1.4.2).

Stevens (1980) and others have considered a variety of techniques that can help to root trees, of which outgroup comparison (including gene duplication events (Iwabe *et al.* 1989)), ontogeny, and paleontology are probably the most useful. The goal in each case is to identify the ancestral condition by some independent means.

Outgroup comparison involves expanding a monophyletic group by one step to include a new taxon which is likely to possess the ancestral state. A nice example of outgroup comparison being used to root a phylogeny is provided by Carson and colleagues' study of 103 species of picture-winged Hawaiian *Drosophila*. (Carson *et al.* 1967; Carson and Kaneshiro 1976; Carson 1983). The unrooted tree was derived from a comparison of chromosome inversions in the different species. *Drosophila primaeva* and *D. attigua*, a pair of sibling species, were treated as the outgroup and used to root the tree—the ancestral node with the most similar inversion patterns to those of *D. primaeva* and *D. attigua* was chosen as the root of the tree. There were two reasons for deciding upon them as the outgroup. First, they possess the chromosome pattern of closely related continental forms which are thought to represent the stock from which the Hawaiian *Drosophila* originated. Second, they are restricted to the oldest island of the Hawaiian chain.

The second method for rooting trees is based on ontogenetic comparisons among species. If evolution of some known characters proceeds by what Gould (1977) calls 'terminal addition', so that derived characters develop later during development than primitive characters, then primitive character states may sometimes be identified using ontogenetic criteria. The value of ontogenetic comparisons to identify primitive states therefore depends on the extent to which a contemporary version of von Baer's (1828) first law holds true: the primitive features of a taxon appear earlier in the embryo than derived features. (von Baer, who was not an evolutionist, would have called them 'general' and 'special' features (Gould 1977)). Such character state changes are, by definition, not reversible and therefore provide information that can help root trees. However, evolution by terminal addition is not a universal phenomenon (see Gould 1977; Kluge and Strauss 1985; Brooks *et al.* 1985*a*, *b*). For example, many cases of neoteny (paedomorphosis) are well documented, which is equivalent to terminal deletion of many characters, and therefore to character state reversal. It might be argued that once complex characters have been lost they will not evolve again (Dollo's Law) but, not surprisingly perhaps, nature even provides exceptions to that generalization. Frogs lost their teeth in the Jurassic, which might be considered the straightforward loss of a fairly complex character. However, the potential to develop teeth seems to have been retained. Indeed, one South American frog *Amphignathodon* has actually re-evolved true teeth in its lower jaw (see Futuyma 1986). Nevertheless, such examples really are exceptions to a useful general rule, and several studies point to the value of comparative ontogenetic sequences in the reconstruction of ancestral character states (e.g. Miyazaki and Mickevich 1982; Alberch and Gale 1983). Although the evidence will often be tentative, if von Baer's first law is appropriately re-phrased and cautiously applied, it is probably 'true enough to be usable' (Ridley 1986*a*, p. 67) in many cases (Fink 1982).

The final technique for helping root trees is to locate appropriate fossil material as corroborative evidence, which may even be used to support the ontogenetic evidence (Miyazaki and Mickevich 1982). However, fossil material can be misleading. For example, taxa with the ancestral character state may be first detected in the fossil record at a later date than those with the derived character state, thus leading to the wrong character state being identified as primitive. That does not, of course, imply that anyone will find a Pre-Cambrian rabbit. Because of the many notorious gaps in the fossil record, there remains considerable debate about its general value as a reliable indicator of the direction of character change (Schaeffer *et al.* 1972; Ghiselin 1972; Paul 1982; Ridley 1986*a*).

3.2.6 On the incompleteness of phylogenetic trees

As we have emphasized, inferring the structure of phylogenetic trees is tricky—we are never sure that we have the correct tree. Comparative biologists should be aware of the fact that they may well be working with the wrong tree! A further problem is that most phylogenies are incomplete in two important ways which mean that comparative methods have to be designed to accommodate the inadequacies (see Section 5.2.1). First, many nodes have more than two daughter branches, and second, many branches which may in fact have been quite different are given the same lengths.

The first problem arises because several consecutive real dichotomies may be compounded into a single multiple branching event. For example, three or more species are often included together in the next higher taxonomic grouping—a single genus. More complete information would enable us to resolve many multiple branch points into a series of dichotomies. However, if speciation involves the simultaneous splitting of one species into three or more daughter species, multiple branch points would be accurate representations within the phylogeny.

The second way in which most phylogenies are incomplete concerns branch lengths and the timing of branch points. Not only are chronological timings usually missing from the tree, but relative timings are artificially classified. For example, when species are classified into genera, families, and orders by some systematists, nodes defining the origins of different families are, by implication, given the same times of occurrence. In theory at least, there is a time earlier than which a node denominates family status, and later a genus. As a consequence, nodes that occurred at very nearly the same time can be assigned a different taxonomic status. Similarly, nodes that occurred at very different times can be placed in the same category. The implicit assumption that equivalent taxa originated at the same time (e.g. the most recent common ancestor of all species in one family existed at the same time as the most recent ancestor of any other family) is made by several comparative methods described in Chapter 5. Of course, finer-grained taxonomic distinctions are often used, and the finer-grained the better as far as comparative biology is concerned. As we shall see in Chapters 4, 5, and 6, the better comparative methods can use any number of taxonomic levels.

3.3 Finding ancestral character states

The problem to be discussed in this section—the second task of this chapter—is how to determine hypothetical ancestral character states given a particular tree structure. As we shall see in Chapters 4 and 5, several types of analysis involve reconstructing hypothetical ancestral character states for the characters being analysed. However, as we discussed in the previous section, we have already considered most of the techniques involved because ancestral character state reconstruction often plays a major part in procedures used for reconstructing phylogenies.

Outgroup comparison has frequently been advocated as a method for reconstructing ancestral character states for dichotomously varying characters under the assumption of parsimony. Take the example of attempting to determine the character state of a common ancestor of a monophyletic group consisting of two species with different character states. One possibility is to expand the monophyletic group by one step to include the minimum number of new species with a less recent common ancestor. If all the additional species (the outgroup) demonstrate a single character state, then parsimony suggests that it is the ancestral condition for the original group (the ingroup). If the common ancestor did not contain the character state of the outgroup, at least one additional evolutionary transition would be required to account for the observed species diversity. If there were more than two species in the original ingroup, one possibility would be to decide on the most common character state among those species as being ancestral. Unfortunately, ingroup comparisons frequently do not provide such parsimonious trees as do outgroup comparisons, partly because it is entirely possible for the most common character in a group to be the derived state (Crisci and Stuessy 1980; Ridley 1983*a*).

How, in practice, is outgroup comparison used? Futuyma (1986) gives a straightforward example. The anterior legs of species belonging to two butterfly families (Nymphalidae and Danaidae) are very reduced in size, whereas the anterior legs of species belonging to two other families (Papilionidae and Pieridae) are functional. Moths and other insect orders, taken as outgroups, have functional anterior legs, which would thereby be accepted as the hypothesized ancestral state. The families with reduced legs would, under outgroup comparison, be considered to share a more recent common ancestor than either does with members of the two families containing functional legs only.

However, outgroup comparison has a serious shortcoming for allocating ancestral character states to nodes: it may provide the most parsimonious *local* number of character state transitions, but not the single most parsimonious *global* number of state transitions (an improved method is provided by Maddison *et al.* 1984, see Box 3.1). If phylogenetic trees are known, algorithms, simple rules, and computer programs have been developed to determine hypothetical ancestral character states based on the parsimony assumption of the fewest number of character state transitions[8].

When characters are continuously varying with no restraints on direction of change, two methods are available to estimate the hypothetical

---

[8] Farris 1970; Fitch 1971; Hartigan 1973; Moore *et al.* 1973; Sankoff and Rousseau 1975; Sankoff and Cedergren 1983; Maddison *et al.* 1984; Swofford and Berlocher 1987; Swofford and Maddison 1987.

Box 3.1. Outgroup comparison, parsimony, and the reconstruction of ancestral character states

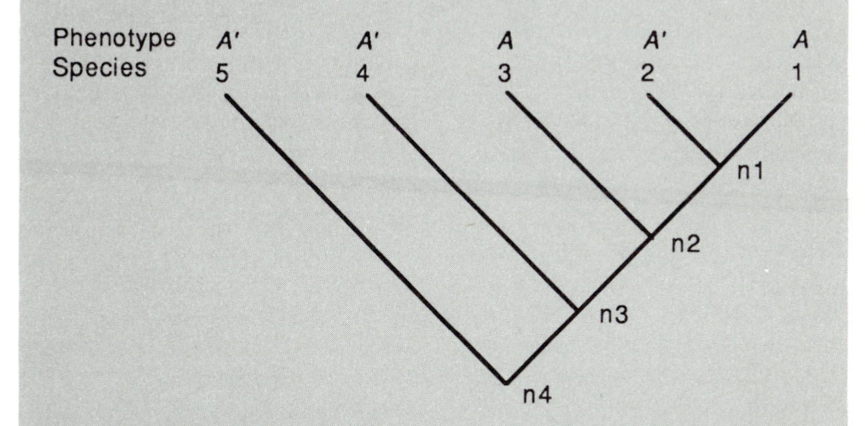

The problem is to allocate ancestral character states to the nodes n1 through n4. The first node, n1, is ambiguous but reference to its outgroup, species 3, suggests that the most parsimonious solution is for nodes n1 and n2 to be labelled A. A single transition from $A \rightarrow A'$ is then required from the node linking n1 with species 2. However, such a solution is only locally parsimonious because, if species 4 is considered, then two equally parsimonious allocations of ancestral character states are possible, one with n3, n2 and n1 labelled A' and the other with the same nodes labelled A. Consideration of species 5 labels both n3 and n4 as A'.

Maddison *et al.* (1984) provide an algorithm for determining the maximally parsimonious ancestral character states from comparison with local outgroups, under the assumption that all branches are of equal length and the probability of the transition from $A \rightarrow A'$ is the same as $A' \rightarrow A$. The algorithm gives rise to two straightforward rules if outgroups consist of single species or monomorphic taxa. The rules also provide satisfactory resolutions when multi-state characters are involved.

Rule One states that, as we move from the unresolved ingroup (species 1 and 2) out, if the first outgroup and the first doublet (consecutive outgroups with matching characters) have the same character state, then that state is the most parsimonious assignment for n1. (The first outgroup may constitute one of the members of the doublet.) When the first outgroup and the first doublet have different character states, no single most parsimonious allocation of ancestral character states exists. Rule Two, which is applied to cases where there are no doublets, states that if the first and last outgroups have the same character state, that state is the most parsimonious assignment for n1, but if the first and last outgroups have different character states, no single most parsimonious allocation of ancestral character states exists.

The trees numbered 1 to 4 below, adapted from Maddison *et al.* (1984), demonstrate the use of the rules in practice. The extreme right hand node in each tree may be *A* or *A'*. We need Rules One and Two to determine the ancestral states at the nodes marked ■.

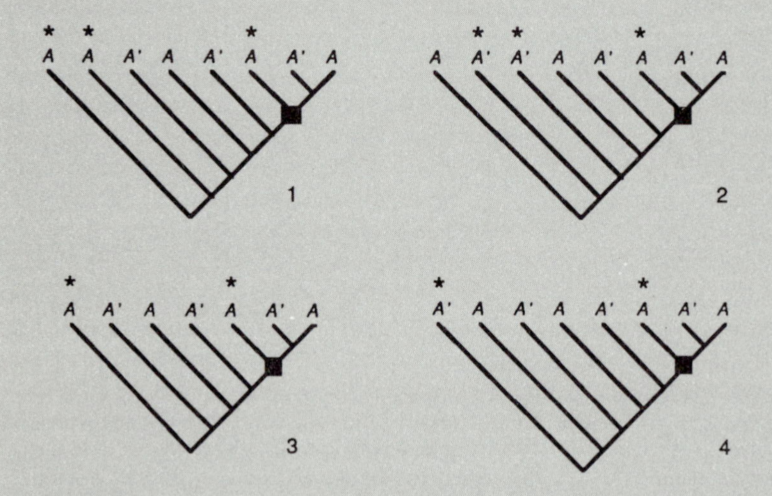

For tree 1, Rule One is applied because there is a doublet. The first outgroup and the first doublet (*) have the matching character state *A*, which is therefore the single most parsimonious ancestral state. Working down through the tree, the ancestral state at each node is *A*.

For tree 2, Rule One is also applied because there is a doublet. The first outgroup and the first doublet (*) do not have matching character states, therefore alternative ancestral character states are equally parsimonious. In this example, using the single outgroup method would designate the node as A whereas an equally parsimonious series of ancestral character states can be constructed with the node designated A'.

For tree 3, Rule Two is applied because there is no doublet. The first and last outgroups (*) have matching character state *A*, which is therefore the single most parsimonious character state.

For tree 4, Rule Two is also applied because there is no doublet. The first and last outgroups (*) do not have matching character states, therefore alternative ancestral character states are equally parsimonious.

A real example where these rules were used to estimate ancestral states for lekking and sexual size dimorphism for the phylogeny of grouse and pheasants (Tetraonidae) is summarized in Chapter 4 (Section 4.4.1 and Fig. 4.3).

The important point for the comparative biologist using parsimony to construct ancestral character states is that alternative equally parsimonious solutions are available for trees 2 and 4. In such cases, it is reasonable to examine the consequences of alternative allocations of ancestral character states. For example, Donoghue (1989) examined the evidence for correlated evolution between dioecy and fleshy propagules versus monoecy and wind-dispersed propagules. As we shall see (Section 4.5.2), he compared his results when ancestral character states at equivocal nodes were counted for the null hypothesis of no correlated change with those when equivocal nodes were counted against the hypothesis.

Outgroup comparison as well as the parsimony algorithm discussed above assume that character change in one direction is as likely as change in the other direction. In fact, once the ancestral character states have been designated, it may be obvious from the resulting allocation of ancestral character states that this assumption is not valid. For example, branches with ancestral nodes A may be less likely to change to A' than vice versa. If $A \rightarrow A'$ is less likely than $A' \rightarrow A$, the algorithm may have underestimated the number of instances of $A' \rightarrow A$. This is an important but largely untackled problem for comparative studies because reconstructing the most likely character states and designating appropriate probabilities of character change are central to testing many evolutionary hypotheses. We shall return to these issues in Chapter 4.

character states at nodes of the most parsimonious tree. The first is the 'median rule algorithm' (Kluge and Farris 1969) which minimizes the summed absolute changes of a character along the branches of a tree by selecting the nodes of the tree from the median value of the three adjacent nodes; an example of its use is given by Larson (1984). The second method, 'the averaging rule algorithm' developed by Felsenstein (in Huey and Bennett 1987; Huey 1987), iterates for each node the average of the three nearest nodes, and minimizes the sum of squared changes along the branches of the tree. The averaging rule method is a modification of Cavalli-Sforza and Edwards' (1964) minimum evolution method which, when applied to blood group data, provided so nice a fit to hypothetical human migration patterns (Cavalli-Sforza *et al.* 1964; Fig. 3.5).

Cavalli-Sforza and Edwards' minimum evolution model applied one of many genetic distance measures to gene frequency data. As we have seen, the minimum evolution criterion is not based directly on any realistic model for evolutionary change. Indeed, Cavalli-Sforza and Edwards (1967, p. 240) themselves emphasized, 'It certainly cannot be justified on the grounds that evolution *proceeds* according to some minimum principle'

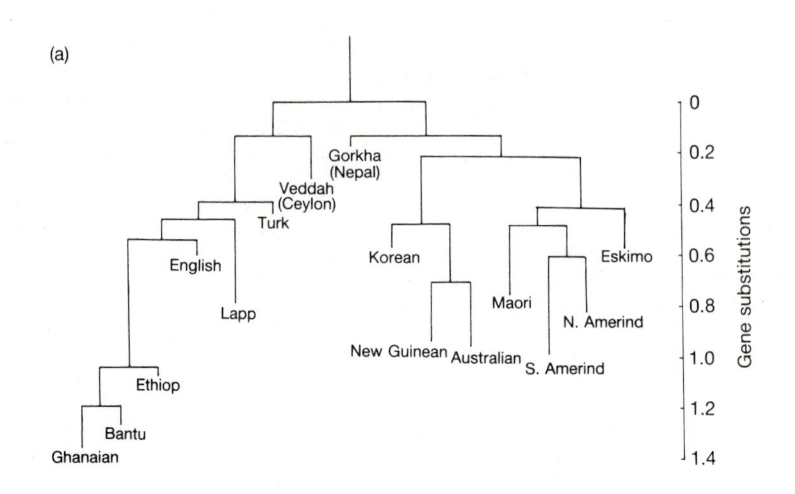

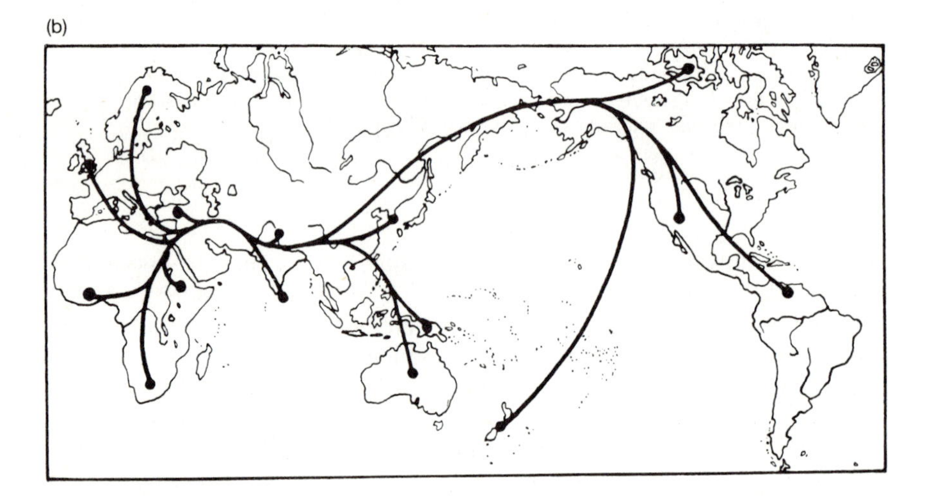

Fig. 3.5. Minimum evolution method used on blood group data from different human populations to produce a gene substitution tree (a), which was then superimposed on a map of the world (b) to show its correspondence with probable historical migration patterns. (After Cavalli-Sforza *et al*. 1964). Subsequent studies have improved on these data.

(their italics). Felsenstein (1985*b*) used a model of evolution for small populations under drift to evaluate several genetic distance measures, and found faults with each of them when initial gene frequencies are extreme. The important point about this exercise is not that Felsenstein's particular model is likely to be realistic, but simply that we need to know how measures of different genetic distance behave under specified conditions. Other comparisons of genetic distance measures under different regimes of evolutionary change (e.g. Latter 1973*a*, *b*; Nei 1976; Reynolds *et al.* 1983), including a neutral mutation model, help towards that end (Felsenstein 1985*b*). But all of this begs the central question: what is the most appropriate model of evolution? And that, perhaps, as the main message of this chapter, provides an appropriate note on which to finish this discussion.

A useful review of available computer programs or packages of programs that can use a variety of specified criteria to reconstruct both phylogenetic trees (including PHYLIP, PAUP and Hennig86) and ancestral character states for characters with specified rules of change (MacClade) can be found in Maddison and Maddison (1989).

3.4 Summary

Comparative methods need to utilize information on phylogenetic tree structure and ancestral character states. Maximum likelihood procedures based on appropriate models of evolution provide one suitable statistical technique for providing that information. Cladistic approaches using parsimony and compatibility criteria can produce approximations to maximum likelihood solutions under some models of evolutionary change. Standard taxonomies are often unsuitable for comparative studies because they do not accurately represent phylogenetic relationships. DNA sequence data from parts of the genome with appropriate amounts of divergence for the taxa being compared can provide particularly useful material for phylogenetic tree reconstruction.

4

Comparative analysis of discrete data

'Few methods have been proposed which are framed explicitly in terms of character change along the lineages of a phylogeny' (Maddison 1990, p. 539).

4.1 Introduction

In this chapter we consider methods for the comparative analysis of discrete characters, or characters that can take only a finite number of states. The methods that we review and develop address the analysis of dichotomous characters—that is, characters that are found in just two states. However, this does not impose any limitations because any multi-category discrete variable can always be represented as a set of dichotomous classifications, each one representing the presence or absence of a particular state. In each branch of the phylogenetic tree, a trait that is initially absent may evolve or a trait that is initially present may disappear. These changes along the branches are called character state transitions or just transitions. The problem, then, is to determine whether a particular pattern of observed transitions between alternative states of two phenotypic characters throughout a phylogenetic tree represents correlated evolutionary change in the two characters, or just a chance pattern of association.

The idea of a variable alternating between one of two character states finds striking parallels in other fields of biology. Island biogeographers have long confronted the problem of species turnover: species initially absent on an island may at a later time be present as a consequence of immigration, and species initially present may be absent as a result of local extinction. In fact, some of the models that we shall apply to characterize character transitions in a phylogeny are in some cases similar to those used by island biogeographers (e.g. MacArthur and Wilson 1967), and in other cases identical (e.g. Diamond and May 1977, note 7). Similar models have also been employed in the genetics literature to examine the consequences of both reversible mutation (e.g. Kempthorne 1957), and various selection processes (e.g. Felsenstein 1979).

Comparative questions involving discrete variables are widespread in biology. Recent studies range over issues from breeding system and mode of dispersal in plants; group living, coloration, and palatability in insects; sexual selection and breeding systems in birds; and sexual competition and social organization in primates (e.g. Givnish 1980; Harvey and Paxton 1981; Ridley 1983*a*, 1986*b*; Sillén-Tullberg 1988; Donoghue 1989; Höglund 1989; Maddison 1990). In Chapter 2 we considered the various biological reasons for closely related species having similar phenotypes. Whenever such similarity is inherited by descent from a common ancestor, we are not justified in treating species as independent points for statistical analysis. Instead, we must identify separate or independent evolutionary origins of the character states of interest. Because this chapter deals with statistical methods for analysing discrete character data obtained from a hierarchically nested phylogeny, we start by presenting a statistical complement to the material contained in Chapter 2. How do the biological processes described in Chapter 2 translate into phylogenetic similarity among discrete character states? And how might this similarity be dealt with? Many of these same themes will recur in Chapter 5 in conjunction with methods of analysis for continuous variables.

We then describe the kinds of transitions between character states that will be observed between two dichotomous discrete variables, and show how two methods designed to analyse discrete data from phylogenies, one developed by Ridley (1983*a*) and the other by Maddison (1990), treat these transitions. This is followed by examples of the application of both methods. Finally, we describe a general statistical method for treating comparative dichotomous data (a method developed by Pagel from one given by Pagel and Harvey 1989*b*). Here the parallels with models in island biogeography become apparent. This method makes it possible to use Ridley's and Maddison's methods while taking into account times of phylogenetic divergence.

4.2 The problem

4.2.1 Non-independence of data points

Suppose that we have collected together data on whether the larvae of each of many species of butterfly are warningly coloured or cryptic, and whether they are solitary or gregarious. We might ask whether there is evidence that warning coloration and gregariousness are related. Two dichotomous variables can give rise to four possible combinations of character states. It is tempting merely to tally the number of species that are in each of the four possible combinations of the characters and perform a statistical test of association, such as the chi-square. But most statistical tests assume that the individual data points are independent, and the proposed chi-square test for association would be no exception. Therefore, for this test to give us correct probability values, we must assume that the phylogenetic structure underlying the evolution of these species allows them to be independent. Felsenstein (1985*a*) gives just such an example for continuous variables, which we shall use again in Chapter 5 and have adopted here for discrete variables (Fig. 4.1).

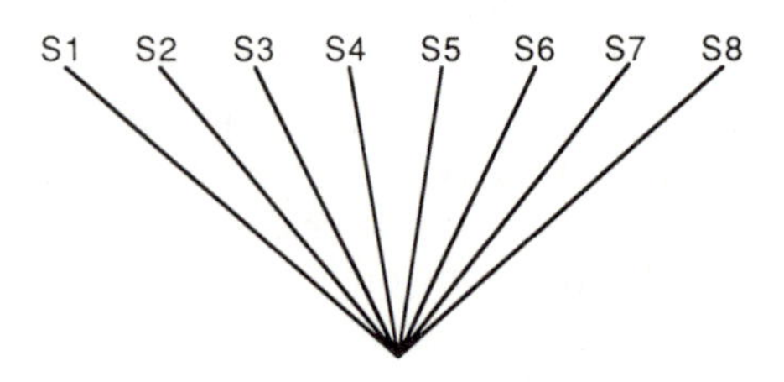

Fig. 4.1. Simultaneous radiation of eight species from a single common ancestor.

The eight species in Fig. 4.1 are all assumed to have radiated at the same time from a single common ancestor. On the assumptions that selective forces are not correlated among lineages and that the underlying genetical structure does not constrain evolution, the changes in any one lineage are independent of the changes in the other lineages, and so species can be treated as independent data points in an analysis. However, consider the more realistic phylogeny shown in Fig. 4.2.

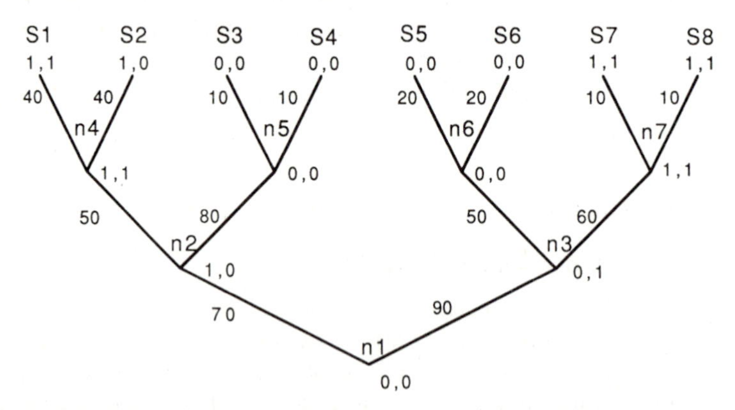

Fig. 4.2. Eight extant species (S1 to S8) which have evolved by a dichotomous branching process. Character states for two discrete variables (X, Y) are given for each species, as well as for ancestral species at each node (n1 to n7). The numbers given at the left of each branch represent times between successive nodes. Branch lengths are not drawn to scale.

We should be reluctant to treat species from this phylogeny as independent. Species S3 and S4, for example, share all of their evolutionary history except for the relatively small amount of independent evolution since they split from their common ancestor: the evolutionary changes leading to their contemporary character states have not been independent. Thus, to count both of them in a statistical test would overestimate the number of times that their particular combination of characters has evolved.

The usual symptom of non-independence is that closely related species tend to be more alike than more distantly related species. This point, of similarity associated with recent shared ancestry, is grasped intuitively, but what evolutionary models are implicit in it?

4.2.2 Two views of how non-independence arises

One view sees similarity associated with phylogeny arising essentially out of a constraint on how fast species can evolve (similar to 'phylogenetic time lags' in Chapter 2). For each of the branches in Fig. 4.2 we can write a variable t_{ij} indicating the length (in units of time or numbers of generations) of the branch connecting node i with node j. If we know the probability of a character changing state per unit of time and we assume that rates of change do not vary, then it is fairly easy to see that unless the probability is high or the branch length is very long, species with very recent common ancestors are more likely to have similar character states than more distantly related species. Thus, for example, consider that the branches leading from nodes n4, n5, n6, and n7 to their respective species are each of length 1.0 instead of the lengths shown on Fig. 4.2. Further, assume that the probability of a particular character changing state per unit time is less than 0.5. Then, more often than not, the species-pairs in the terminal branches will have the same character state, but not because they have each independently evolved it.

There is no doubt an upper bound to the rate of evolution, but whether it is important for the characters typically studied in comparative analyses is less certain. For some characters it probably is, others not. For example, the odds against a marsupial mammal evolving, in a single speciation, into a placental mammal are astronomical. Examples such as this may suffer from adopting the wrong level of analysis. That is, there are very likely intermediate degrees of placentation that can vary among species in, for example, the same genus. But the deeper problem with the 'not enough time' point of view is that it suggests that the only reason for phylogenetic similarity is that species have not spent enough time apart to become very different. This, surely, is not the whole story.

The second point of view, and the one implicit in the adaptationist explanations that we termed 'phylogenetic niche conservatism' and 'different adaptive responses' in Chapter 2, is that phylogenetic similarity reflects a similarity of selective forces and a similarity of response to similar forces. On this view, two species in the same genus are similar because they have shared histories of selective forces up to the point of their splitting, and because they are very likely to share many features in their current environments. The latter point is virtually guaranteed if we believe that similar species tend to be adapted to similar environments. It does not require any special models of speciation.

4.2.3 Implications of non-independence for statistical tests

Either viewpoint on phylogenetic similarity has important implications for testing comparative hypotheses. Species may not be independent with respect to some character because they have not, even in spite of selection to diverge, had enough time to become different. Non-independence will also arise due to shared selective forces and similar phenotypic responses to those forces. Correlating the character states of two variables across species, then, could include many data points that were not independent. As a consequence, the degrees of freedom for the statistical test will be overestimated and we cannot trust the probability values associated with the statistic.

We could, alternatively, restrict ourselves to higher nodes of the tree on the grounds that they are more likely to be independent. But this will be palliative at best, because exactly the same logic that was applied to the species can also be applied to differences among the higher nodes. The use of higher nodes rests implicitly on the assumptions of long enough branches or high enough rates of evolutionary change (or both), or such very different selective forces among the lineages that the higher nodes can be considered independent with respect to the characters under consideration.

4.3 Character transitions and discrete variables

If we often cannot consider species or even higher nodes as independent, then how can we use the information in the phylogeny to test for covariation between the two characters? One answer was hinted at in Fig. 4.1. The subtaxa within a monophyletic group share their evolutionary history up to the point of their origins. But, as a first approximation, the evolutionary *change* along an individual branch is independent of the change in other branches. Thus, one way to test whether two discrete variables have evolved in a correlated fashion is simply to correlate the changes in X and Y throughout a phylogeny. For example, between nodes n1 and n2 of Fig. 4.2, one variable changed from state 0 to state 1 but the other did not, which would count against the idea of correlated change between characters. Continuing up the tree we could tally all such changes, ignoring branches in which neither character changes, and find their correlation. This is very close to what the method developed and used by Ridley does (Ridley 1983a, 1986b, 1988). We shall return to Ridley's method later on. But for now it is important only to realize that counting changes in different branches of the tree avoids the problem of non-independence from either point of view expressed above.

With two dichotomous characters there are, in fact, 16 possible combinations of change and absence of change that can take place along the branches of a phylogenetic tree. For example, consider two characters, X and Y, that can each exist in state 0 or 1. Possible states at the beginning and end of a branch for each character are 0→0, 0→1, 1→0, and 1→1. Each pair of states for one character can be associated with any of the four possible pairings for the other character, making a total of 16 combinations which we list in Table 4.1 for future reference. Nine of the combinations are evident along the 14 branches of Fig. 4.2: one combination occurs four times, one three times, and the other seven once each. In the following sections, we describe Ridley's (1983*a*) and Maddison's (1990) methods in terms of these transitions.

Table 4.1 Characters X and Y can each exist in state 0 or 1. There are 16 possible types of branch in a phylogenetic tree classified according to beginning and end states of the two characters. The numbered classes are those referred to in Table 4.3 and subsequently in the text.

	Transitions in character Y			
Transitions in character X	0→1	0→0	1→0	1→1
0→1	1	2	3	4
0→0	5	6	7	8
1→0	9	10	11	12
1→1	13	14	15	16

4.4 Counting evolutionary events

Ridley (1983*a*) designed his method for counting evolutionary events to get around the problems of non-independence in phylogenies. The method begins with a phylogeny that has been reconstructed according to one of the methods that was described in Chapter 3. It is assumed that the reconstructed phylogeny represents our best estimate of the true phylogeny. To assign character states to higher nodes, Ridley suggests using outgroup analysis. However, as described in Chapter 3, rules suggested by Maddison *et al.* (1984) and by Swofford and Maddison (1987) are more appropriate if parsimony is the criterion to be used for reconstructing ancestral character states.

Once the higher nodes have been assigned, one works through the phylogeny, keeping a tally of the number of transitions in the tree (i.e. branches for which the beginning and end-states differ) that have *ended* in each of the four possible combinations of end-states. A transition is defined as a change of state in either or both characters along a branch. Because only the states of characters at the end of each branch are tallied, there will be only four different combinations. Branches along which no change occurs are not included. By not including such branches, the method avoids counting species or higher nodes which share character states with an immediate common ancestor, and which thereby cannot be considered independent data points.

The result of the tallying is a two-by-two contingency table, showing the number of branches along which transitions occurred that ended in each of the four states. The hypothesis that changes in X and Y are independent is tested by means of a chi-square statistic. A significant Chi-square (or Fisher's Exact Test) is evidence that the transitions have not been independent; some combinations of characters are more (or less) common than expected by chance. Table 4.2 shows the counts obtained by applying Ridley's method to the data in Fig. 4.2.

Table 4.2 Analysis of evolutionary events using Ridley's method applied to the data from Fig. 4.2. A tally of end-states in branches showing changes in X or Y produces the 2 × 2 contingency table, for which χ^2=0.19, ϕ= 0.17, and P>0.50.

	Y=0	Y=1
X=0	2	1
X=1	2	2

The logic of Ridley's method can be understood in terms of the 16 phylogenetic tree branch types shown in Table 4.1. Four of the 16 types (no change in either character) are ignored by Ridley, whereas the other 12 are treated together in four groups of three depending on the end-states. Thus, for example, branches with characters *X* and *Y* both in end-state 0 can arise from three types of transition (*X* changing from state 1, *Y* changing from state 1, or both characters changing from state 1). Ridley's method treats the three different combinations as equivalent. In Table 4.3 we show, by reference to Table 4.1, which transitions are classified together by Ridley's method.

Table 4.3 The way in which transitions from Table 4.1 contribute to Ridley's method of analysis. Types 6, 8, 14, and 16 are not included because they describe branches along which no transitions occur.

	Y=0	Y=1
X=0	7, 10, 11	5, 9, 12
X=1	2, 3, 15	1, 4, 13

Ridley's method, then, is correctly seen to be a test of whether end-states, defined as the character states at the end of a branch, tend to be correlated. This interpretation suggests an imaginary 'phylogeny' consisting of many parallel branches. The beginning states of the branches are unimportant provided that one or the other of the characters has changed along the branch. If changes in the two characters are random, then there should not be an association among the end-states. Correlated evolutionary change, however, will produce a correlation among the end-states. The method does not utilize the direction of evolutionary change, or which variable changed first. These will be key differences between Ridley's and Maddison's approaches (Section 4.5).

4.4.1 Applying Ridley's method: lekking and dimorphism

An example of the use of Ridley's method comes from Höglund's study of lekking and sexual dimorphism in size and plumage among birds. Höglund (1989) found a significant association between lekking and both size and plumage dimorphism across 114 bird species: lekking was associated with a higher proportion of the species showing size dimorphism and colour dimorphism (Table 4.4).

However, Höglund questioned these results on the grounds that all three variables show strong phylogenetic associations. To determine whether a relationship between lekking and dimorphism has actually arisen through the correlated evolution of the two characters, Höglund constructed phylogenies for his species, determined likely ancestral character states,

Table 4.4 Lekking and sexual dimorphism in size and plumage in birds. Each data point represents a single species. The associations between lekking and size and plumage dimorphism are both significant at P<0.001 in this comparison with species counted as independent points. (After Höglund 1989).

Mating system	Sexual size dimorphism		Plumage dimorphism	
	Present	Absent	Present	Absent
Lekking	69	18	55	33
Non-lekking	13	13	8	18

and counted the number of independent instances of the evolution of lekking and of sexual dimorphism in size and plumage. Fig. 4.3 shows the phylogeny for the Tetraonidae (grouse and pheasants). The phylogeny illustrates many of the principles of reconstructing ancestral character states, as well as the steps for conducting Ridley's suggested type of analysis.

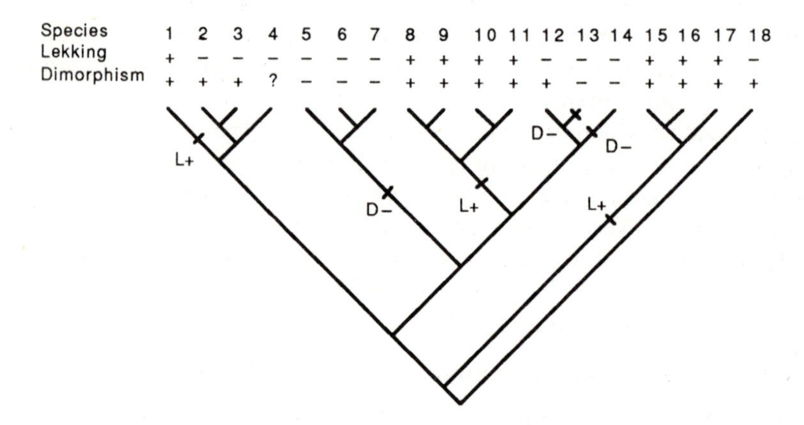

Fig. 4.3. Lekking and sexual size dimorphism in the Tetraonidae. All species in this family are sexually dimorphic for plumage. Transitions on the phylogenetic tree (L, lekking; D, dimorphism; +, gain; −, loss) are calculated using the rules given by Maddison *et al.* (1984), described in Chapter 3 (Box 3.1). Species referred to by number are: (1) *Centrocercus urophasianus;* (2) *Dendragapus falcipennis;* (3) *D. canadensis;* (4) *D. obscurus;* (5) *Lagopus leucurus;* (6) *L. mutus;* (7) *L. lagopus;* (8) *Tetrao urogallus;* (9) *T. parvirostris;* (10) *T. textrix;* (11) *T. mlokosiewiczi;* (12) *Bonasa sewerzowi;* (13) *B. bonasia;* (14) *B. umbellus;* (15) *Tympanuchus cupido;* (16) *T. pallidicinctus;* (17) *T. phasianellus;* (18) Phasianinae (outgroup). (After Höglund 1989).

Höglund's concerns about phylogenetic similarity are borne out: all members of the family Tetraonidae are dimorphic for plumage and yet each of the species is counted as a separate value in the test given in Table 4.4. Focusing on the other two variables, Höglund inferred that sexual dimorphism in size was lost at least three times while lekking appeared at least three times during the evolution of the Tetraonidae. Each of the instances of the loss of size dimorphism occurred in a non-lekking species and each instance of the gain of lekking occurred in a size-dimorphic species, a perfect association within this family.

However, the patterns were not as strong in other families. Table 4.5 shows the distribution of evolutionary changes for all the families combined. The result, in contrast to the highly significant across-species analysis, is not significant. Höglund's careful phylogenetic analysis reveals no significant evolutionary association between dimorphism in size and plumage in lekking birds. These results could change as new phylogenetic data become available, but they stand as a caution against drawing evolutionary inferences from analyses performed across species. Höglund also suggests that an analysis of size or plumage dimorphism may overlook other characters that females use in mate choice. For example, in the absence of plumage dimorphism, song dimorphism may have evolved.

Table 4.5 Lekking and sexual dimorphism in size and plumage in birds. The data points refer to independent comparisons calculated using Ridley's method. Using Fisher's Exact Test, neither association is significant. (After Höglund 1989).

Mating system	Sexual size dimorphism		Plumage dimorphism	
	Present	Absent	Present	Absent
Lekking	11	9	11	8
Non-lekking	6	9	6	7

Ridley has applied his method to a variety of cases where correlated evolution between characters might be expected (Ridley 1983*a*, 1986*a*, 1988). For example, he found that precopulatory mate guarding by male invertebrates tends to be evolutionarily associated with moulting by females, presumably because sexual receptivity predictably follows moulting (Ridley 1983*a*). And in a phylogenetically more diverse sample ranging from flies to fishes, Ridley (1983*a*) found that animals tend to mate with partners that are the same size as themselves among species in which: (1)

larger females lay more eggs; (2) larger males have an advantage in competition for mates; and (3) the duration of mating is long.

4.5 Tests of directional hypotheses with discrete characters

Ridley's method is designed to detect the pattern but not necessarily the direction of evolutionary change throughout a phylogeny. In this section we describe a test developed by Maddison (1990) for that purpose.

Maddison's test is designed to detect whether changes in one character are concentrated in certain regions of a phylogenetic tree. This might arise if the state of one character somehow makes the evolution of another more likely. Thus, Maddison's test explicitly treats one of the variables as the independent or 'causal' variable and the other as the dependent variable. It then searches for evidence that the likelihood of the dependent variable changing is higher in the presence of one category of the independent variable than in the presence of the other.

An example of an explicitly directional hypothesis of the sort for which Maddison designed his test is given by Sillén-Tullberg's (1988) work on the evolution of gregariousness in butterflies with warningly coloured larvae. Sillén-Tullberg found that gregariousness had evolved 23 times in the butterflies, and in 15 to 18 of these cases, the larvae were warningly coloured. Sillén-Tullberg concluded that this many gains of gregariousness in the presence of warning coloration suggest that warning coloration predisposes butterflies to evolve gregariousness.

Maddison's (1990) criticism of Sillén-Tullberg's approach, however, is that it fails to take into account the distribution of warning coloration in the phylogeny. Maddison points out that if, for example, warning coloration is very common in the butterflies then we would expect gregariousness to evolve in its presence more often than not, just by chance. This is easy to see in an extreme case: if all butterflies are warningly coloured then any evolutionary change toward gregariousness will be in the presence of warning coloration. Maddison's test, then, attempts to take into account the phylogenetic distribution of the traits, in addition to the number and pattern of transitions, in deciding whether a particular pattern is evidence for correlated evolution.

To see how Maddison takes into account the phylogenetic distribution of traits, consider the simple phylogeny in Fig. 4.4.

Two characters are coded on the tree. One, the independent variable can take the states *W* and *B*. The other takes the values 0 and 1. We take the ancestral condition for this tree to be *W*0, and thus the tree shows two instances of the evolution of *B* and two instances of the evolution of 1. To keep this example simple, we assume perfect knowledge of the phylogeny,

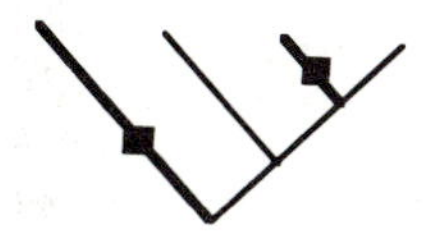

Fig. 4.4. A hypothetical phylogenetic tree. Branches marked in bold refer to lineages in which a character is in state *B*, with other branches in state *W*. Diamonds crossing the bold branches refer to evolutionary transitions of a second character from ancestral state 0 to derived state 1.

and that the 1s have evolved in both cases after the *B*s. How should we test whether the 1s are more likely to evolve in the presence of *B* than in the presence of *W*?

We want to test the likelihood of having two gains of 1 in the presence of *B*, given the distribution of *W* and *B* throughout the tree. To see why, it is important to take into account the distribution of *W* and *B*. Consider that there is only one way in Fig. 4.4 that both 1s could have evolved in the presence of *B*, but many ways that they could have evolved if *W* and *B* are ignored. This is the essence of Maddison's test: we must calculate (*a*) the number of different ways that there are to have the two gains of 1 in the *B* branches of the tree given that there are two gains total (or more generally, the number of different ways that there are to have x or more gains of 1 in the *B* branches of the tree given that there are y gains of 1 and z losses of 1 over the whole tree: $x \leqslant y$); and (*b*) the number of ways that there are to have two gains of 1 on the tree without regard to whether they occur in *W* or *B* branches (or more generally, the number of different ways that there are to have y gains and z losses on the tree without regard to whether they occur in the *W* or in the *B* branches). The probability of the particular result under the null hypothesis is a/b. The denominator, then, records the number of ways that the pattern of gains or losses in the dependent variable could have occurred on the tree. The numerator is a subset of this number. It records the number of ways that a pattern of transitions as extreme or more extreme than that observed could have occurred in branches of the tree in which the independent variable takes one particular state (e.g. *B*).

Maddison's test can be illustrated for the tree in Fig. 4.4. It is straightforward to enumerate all possible outcomes corresponding to the quantities *a* and *b*. There turn out to be nine different ways that character state 1 can evolve twice on the tree (Box 4.1). Only one of these corresponds to evolving both 1s along the *B* branches of the tree. Thus the test yields a probability of 1/9 by chance alone that both instances of the evolution of '1' would occur in the *B* branches of the tree.

Box 4.1. The nine possible ways that 1 can evolve twice on the phylogenetic tree in Fig. 4.4 (top tree here)

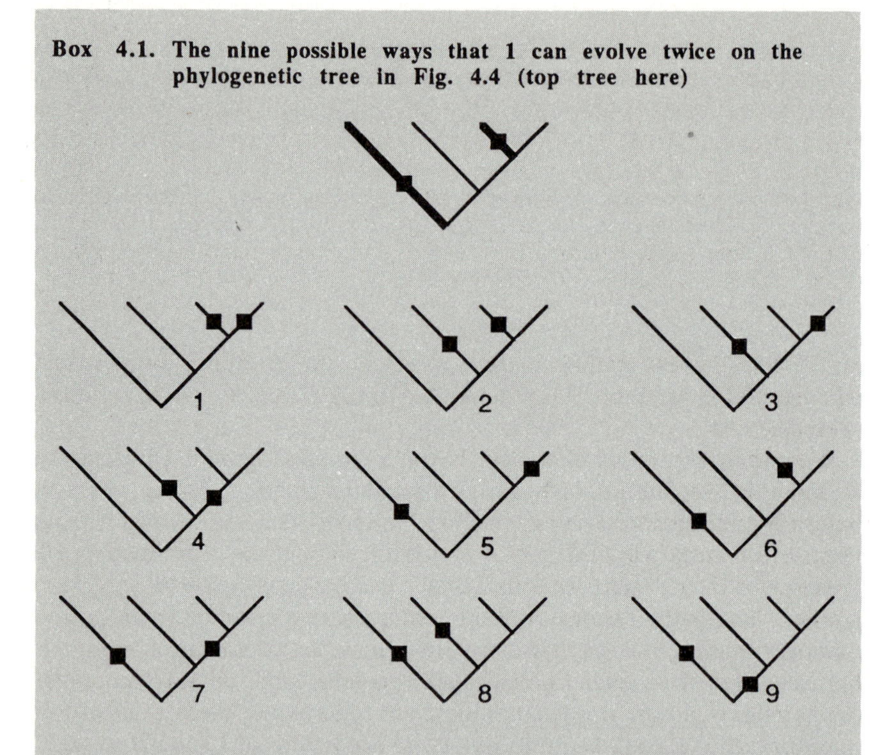

The nine redistributions of character states exhaust the possible ways of evolving the character 1 twice on the tree. Only one of them corresponds to the character evolving both times in the bold branches of the tree. Thus Maddison's test would assign a probability 1/9 to the likelihood that both $0 \rightarrow 1$ transitions occurred on bold branches by chance alone.

It might be thought that a much simpler test of this hypothesis would be to compare the proportion of *B* branches in which 1 has evolved with the proportion of *W* branches in which 1 has evolved. Maddison (1990) correctly points out that such a test ignores the fact that the topology of trees affects the number of ways that events can occur. This is most easily seen with an example. Box 4.2 shows a tree that, like the tree in Fig. 4.4, has two instances of the evolution of 1 in the presence of *B*, no losses of 1, and four *W* and two *B* branches. However, there are 11 possible ways that two instances of 1 could have evolved on the tree in Box 4.2 compared to the nine for the tree in Fig. 4.4 (Box 4.1). Thus, identical patterns of 1s evolving on the *B* branches of these two trees nevertheless have different probabilities of occurring by chance.

Box 4.2. How tree topology can influence the likelihood of correlated evolutionary events

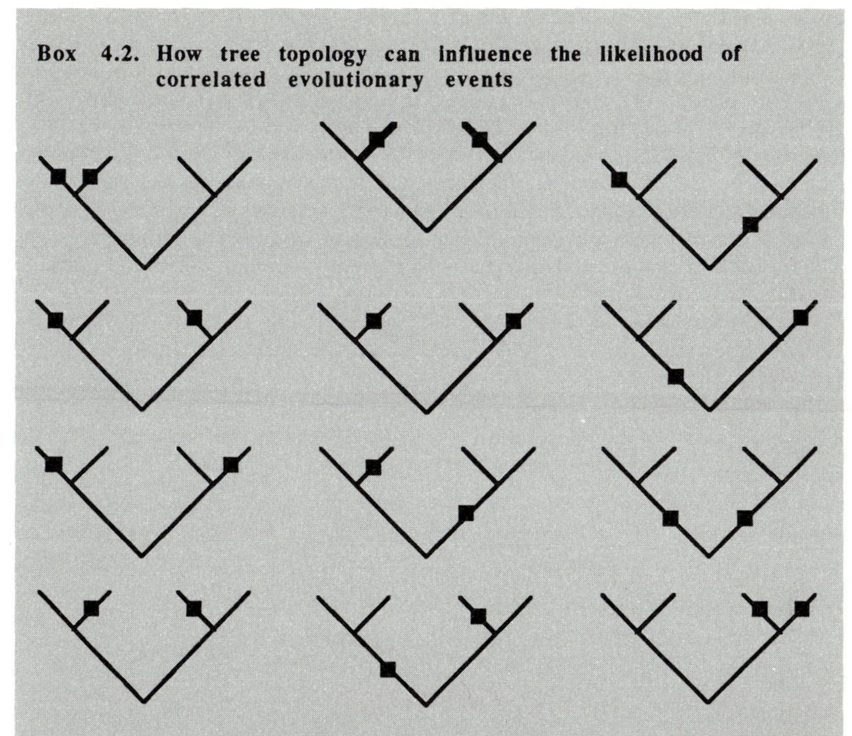

The top centre phylogenetic tree has several similarities with the tree shown in Fig. 4.4: it has the same number of branches in states *B* (bold) and *W*, the same number of character state transitions from 0 to 1, and both transitions occur on the *B* branches. However, the 11 ways that the character transition $0 \to 1$ can occur (as enumerated by the trees drawn next to and below the top centre tree) contrast with the nine arrangements for the tree shown in Fig. 4.4 (see Box 4.1).

For simple trees, Maddison gives an algorithm for computing the two values required for his test. However, this algorithm quickly becomes impractical to use for complex trees and so Maddison has developed a simulation technique for such cases. Maddison re-analysed Sillén-Tullberg's data set with this method and found that 15 to 18 gains of gregariousness in the presence of warning coloration could easily have occurred by chance, given that warning coloration is so widespread in the butterflies (Table 4.6 and Fig. 4.5).

4.5.1 Transitions among character states

Maddison's method uses the 16 transitions in Table 4.1 differently from Ridley's method. Returning to the example of Höglund's (1989) study of

Table 4.6 Data from Sillén-Tullberg's (1988) study of warning coloration (aposematism) and gregariousness in butterflies. Each record is of an evolutionary event. There are 15 to 18 instances of the evolution of gregariousness following the evolution of warning coloration. Using Sillén-Tullberg's estimates (see text): probability of 15 or more out of 20 (two-tailed test) = 0.04, probability of 15 or more out of 23 = 0.21, and probability of 18 or more out of 23 = 0.01. Families included are the Papilionidae, Pieridae, and the Nymphalidae. Re-analysis by Maddison's (1990) method, based on 23 gains and six losses of gregariousness and taking into account the phylogenetic distribution of warning coloration, gives P> 0.05 for both 15 or more and 18 or more gains in the warningly coloured branches. (See Fig. 4.5).

State of larvae	Solitary	Gregarious
Evolution of aposematism	9	0
State of larvae	Aposematic	Cryptic
Evolution of gregariousness	15	5
Evolution of aposematism and gregariousness inseparable	3	

lekking and sexual dimorphism, we showed how Ridley's method counted any transition in the phylogeny that resulted in either lekking and dimorphism or non-lekking and monomorphism as evidence for the relationship. But, for example, there are three ways that an end-state of lekking and dimorphism can arise: (1) a monomorphic lekking taxon may become dimorphic; (2) a dimorphic non-lekking taxon may evolve lekking; or (3) a monomorphic non-lekking taxon may show a double transition to dimorphism and lekking. Transitions with qualitatively different causalities are treated equivalently.

Maddison, on the other hand, would use each of these transitions differently. The first would be used as evidence for the directional hypothesis that the evolution of lekking precedes the evolution of dimorphism. The second outcome would be taken as evidence against the hypothesis. To see why, recall that Maddison's test involves counting the number of different ways that, in the present example, the observed

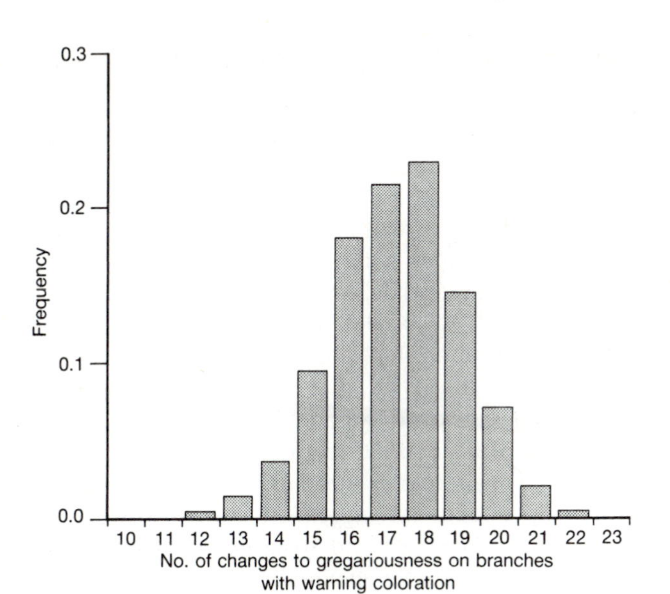

Fig. 4.5. Frequency distribution of the expected number of changes to gregariousness in the warningly coloured branches of Sillén-Tullberg's (1988) phylogenetic tree of butterfly caterpillars, assuming the observed case of 23 gains and six losses in the tree as a whole. The distribution is based on 10 411 computer simulations. More than 20 gains would be judged significant; Sillén-Tullberg observed 15 to 18 gains. (After Maddison 1990).

number of transitions to dimorphism in the lekking branches of the tree plus any other gain of dimorphism (not to mention losses) could have evolved on the tree. The more times that dimorphism has evolved in non-lekking branches, the greater the quantity a in Maddison's test will be, other things equal (e.g. compare the number of different combinations of five things taken five at a time versus five things taken three at a time). Because this number is the numerator of Maddison's test, it will increase the overall probability. The third outcome might be ignored since the order of the changes in a double transition is ambiguous. Alternatively, branches in which both characters change together could be assigned to first one then the other directional category to determine what effect, if any, their inclusion has on the results.

Maddison's method, then, explicitly tests whether the probability of the dependent variable (Y) changing from, say, 0 to 1 depends upon the state of the independent variable (X). This is equivalent to testing whether the probability associated with the cell numbered 5 in Table 4.1 differs from that of cell 13, taking into account back transitions and lack of change.

4.5.2 Maddison's method applied: the evolution of dioecy

Donoghue (1989) used Maddison's method to investigate the evolution of dioecy in flowering plants. Dioecious plants have separate male and female flowers, with only one sex per plant. Monoecious species have bisexual flowers. Givnish (1980) showed a strong correlation across species between dioecy and fleshy propagules that are dispersed by animals, and monoecy and propagules dispersed by wind: 339 of 384 monoecious species had wind-dispersed propagules, while 402 of 420 dioecious species had fleshy propagules ($\chi_1^2 = 570$, $P < 0.00001$). Givnish argued on the basis of this highly significant association that dioecy would be favoured in plants with fleshy propagules.

Donoghue (1989), however, was concerned that strong phylogenetic associations in these traits had led to gross inflation of the strength of the association that Givnish found. Further, Donoghue was interested in Givnish's explicitly directional suggestion that dioecy evolves after the evolution of fleshy propagules, and so used Maddison's approach. The cladogram in Fig. 4.6 shows the evolution of dioecy/monoecy and fleshy/non-fleshy propagules in the Gymnosperms. Ancestral states were constructed according to parsimony rules (Swofford and Maddison 1987). The mode of dispersal was uncertain in some taxa and so Donoghue tested his ideas in two ways using the cladogram in Fig. 4.6: one in which all equivocal taxa were taken to be animal dispersed and one in which they were treated as wind dispersed. Further, some branches showed simultaneous change in both characters. The results did not depend on which way the equivocal taxa were counted and so we confine our remarks only to the effect of the branches in which both characters changed.

When the branches with changes in both characters were treated as evidence for the relationship, Maddison's test gave a significant result: there were five or more gains of dioecy in branches with fleshy propagules, out of seven gains and two losses overall (Fig. 4.6, Table 4.7, $P = 0.02$). However, when the equivocal branches were not counted in favour of the hypothesis, the result became non-significant ($P = 0.13$). Donoghue cautioned that the uncertainties about phylogenies mean that his results cannot be taken as an unambiguous test of Givnish's idea, but rather as a beginning that will encourage further phylogenetic study. Leaving aside such difficulties, this example and Höglund's (1989) lekking example demonstrate how badly we can be misled when species are treated as independent data points.

4.6 A statistical model of evolutionary change

Ridley's and Maddison's methods both account for non-independence of species and provide ways of counting only independent instances of

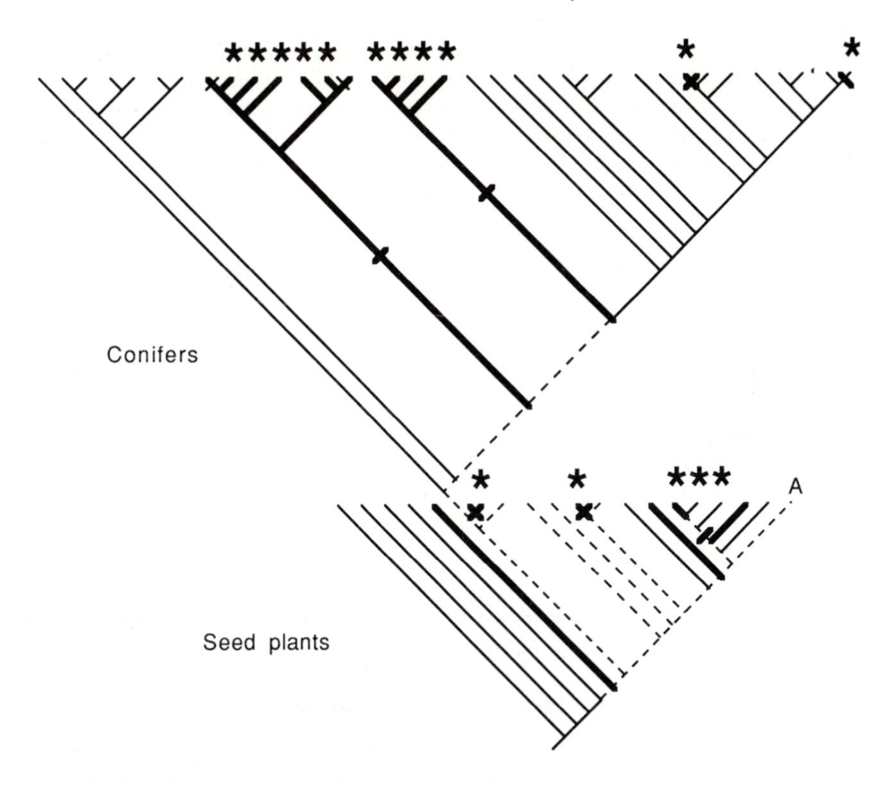

Fig. 4.6. A cladogram of seed plants showing parsimonious reconstructions of the evolution of fleshy propagules (animal dispersal) and dioecy. The phylogeny of angiosperms (A) is omitted from this figure. Branches in bold indicate fleshy propagules, non-bold branches dry propagules, and dashed branches are equivocal. Taxa known to be dioecious are marked by asterisks. Bold crossbars indicate the origins of dioecy and non-bold crossbars reversal to monoecy. (After Donoghue 1989).

evolutionary changes throughout a phylogeny. However, missing from both approaches is any way of taking into account the differing amounts of evolutionary change that might be expected to take place in branches of different lengths. This is an important point, because, for example, we may expect by chance that a variable will be more likely to change in a longer branch of a tree than in a shorter branch. Correlated change in a longer branch may be more likely by the same reasoning. We need a way of correcting for these differences. This section describes a method sketched out by Pagel and Harvey (1989*b*) and formally developed by Pagel (unpublished manuscript) for the evolution of discrete characters that takes into account both evolutionary branch lengths and the probability of character transitions. Ridley's and Maddison's methods can be derived

Table 4.7 Counts of the evolution of dioecy in gymnosperm lineages with fleshy propagules versus wind-dispersed propagules (data from Donoghue 1989). Analysis of this data set using Maddison's method which, despite the small number of changes, by taking into account the number of branches with fleshy and wind-dispersed propagules, yields P=0.02 if equivocal data are counted in favour of the hypothesis, but P=0.13 when equivocal data are not counted.

	Independent variable		
Dependent variable	Fleshy propagules	Wind-dispersed propagules	Equivocal propagules
Gains of dioecy	5	1	1
Losses of dioecy	2	0	0

from this model by making special assumptions about branch lengths and transition probabilities (see Section 4.9).

4.6.1 Evolutionary change in discrete characters

Consider again the hypothetical phylogeny in Fig. 4.2. We shall assume that it is the true phylogeny, and that we know the branch lengths and the character states at higher nodes. Each character can take one of two states, 0 or 1. A branch beginning in 0 may change to 1 in a given branch, or remain in state 0. The same is true for branches beginning in state 1. Now, given a null hypothesis of evolutionary change which treats the two characters under consideration as undergoing independent evolution, we can derive the expected values of the characters and of the variances of character changes in each branch of the phylogeny. The observed states can then be transformed to take into account these expectations. Then, we can correlate the transformed variables to test for covariation in the two characters.

We assume that the evolution of a dichotomous character can be modelled by a process that allows the character to change from one state to the other with specified probability per unit of time. The probability of a change is assumed to be the same in all branches of the phylogeny. For probabilities greater than zero, the model assumes that, over long enough time periods, the character is very likely to change state. Over very short time periods, the character can change but it is less likely to do so. An analogy with base substitutions can be used to illustrate what we mean by a probability of a character change. Assume that a base changes state at some constant rate, r, when measured over long enough time periods. Presumably all of the forces required to cause a base change are present in the base or its environment, but nevertheless the base does not change every $1/r$ units of time, rather there will be a distribution of events (most likely Poisson), with a mean of $1/r$. This distribution says something about the probability of all the forces required for a base change coming together, and the variance of that process. Similarly, for more complex characters we assume that each has the potential to change, but when on average it actually does change depends upon the likelihood of all of the forces for change occurring together or perhaps in the right sequence. Transition probabilities, then, encapsulate our knowledge of all the forces acting on a variable in a particular branch.

More formally, assume that we have a dichotomous variable, X, which can take the values 0 and 1. Define αdt as the probability that X changes from state 0 to state 1 over some small unit of time dt:

$$P_{01}(dt) = \alpha dt. \tag{4.1}$$

Similarly, we can define βdt as the probability of X changing from 1 to 0 over a small unit of time dt:

$$P_{10}(dt) = \beta dt. \tag{4.2}$$

Note that α and β do not have to add to 1.0. For simplicity in the present discussion, we assume that α and β are constants. That is, the probability of a transition does not change in different branches of the tree. However, to make the equations more general one could simply attach subscripts i and j to any transition probability to indicate the two nodes that define the branch in the tree it described. Similar transition probabilities between character states could be defined for a second character Y, and for any number of other variables (the equations describing a second and additional variables will be identical to those for α and β).

The probability of a character X ending in state 1 in a branch of arbitrary length t that began in state 0 can be modelled by a process that allows X to change forward and backward an indefinite number of times. The probability is a function of the relative values of α and β and the length of the branch:

$$P_{01}(t) = \frac{\alpha}{\alpha + \beta}(1 - \exp[-(\alpha + \beta)t]). \tag{4.3}$$

Similarly, if the character begins in state 1, then the probability that after time t it is in state 0 is given by:

$$P_{10}(t) = \frac{\beta}{\alpha + \beta}(1 - \exp[-(\alpha + \beta)t]). \qquad (4.4)$$

Diamond and May (1977, note 7) reported eqns (4.3) and (4.4) in their study of species' turnover rates on islands as a function of immigration and local extinction rates. By the usual methods, we can derive expected values for the mean and variance of the characters. The expected value can be thought of as follows. If evolution was randomly re-run many times, some of the times a branch beginning in state 0 will end in state 1, the remainder of the times it would end in state 0. The average of these many 'trials' is the expected value. The variance is just the variance of these outcomes. The expected character state over an arbitrary amount of time t is given by[9]:

$$\mathrm{E}[X_{01}(t)] = \frac{\alpha}{\alpha + \beta}(1 - \exp[-(\alpha + \beta)t]) = \mu_{01}. \qquad (4.5)$$

The expected variance of the character after time t is given by:

$$\mathrm{E}[\mathrm{VAR}(X_{01})] = \mu_{01}(1 - \mu_{01}). \qquad (4.6)$$

Similar expressions can be derived for the probability of a transition from 1 to 0. Thus, the expected character state after time t for a branch beginning in state 1 is:

$$\mathrm{E}[X_{10}(t)] = 1 - \frac{\beta}{\alpha + \beta}(1 - \exp[-(\alpha + \beta)t]) = \mu_{10}. \qquad (4.7)$$

The expected variance is given by:

$$\mathrm{E}[\mathrm{VAR}(X_{10})] = \mu_{10}(1 - \mu_{10}). \qquad (4.8)$$

The expected values in eqns (4.5) to (4.8) can be used to transform all observations in the phylogeny to have the same mean and variance. Thus, for any given end-state (0 or 1) along a branch that began in 0, a standard

---

[9] Equations (4.5) to (4.8), depend upon the arbitrary assignment of 0 and 1 to the two character states. For different values, the equations will change appropriately and the same ultimate answers will be obtained.

score can be formed by subtracting its mean (eqn 4.5) and dividing by the square root of its variance (eqn 4.6). Similarly, subtracting eqn (4.7) and dividing by the square root of eqn (4.8) will convert to a standard score an end-state that began in state 1. Doing this for each branch will scale all observations to a mean of zero and a standard deviation of one. The size of the standard score (either positive or negative) gives an indication of the degree of change relative to what would have been expected by chance. All the equations for transforming *X* and a second character, *Y*, to standard scores are shown in Box 4.3, along with an example transformation.

The formulae in eqns (4.5) to (4.8) yield positive standardized scores for transitions either from 1 to 0 or from 0 to 1, and negative standardized scores for a lack of change, be it either a branch beginning and ending in 0

Box 4.3. Equations for finding the mean and variance of change along a branch of a phylogenetic tree, and the computation of standard scores

I. Transition probabilities

$$P_{01}(dt) = \alpha dt$$

$$P_{10}(dt) = \beta dt$$

II. Expected character state after time t

$$\mu_{01} = \frac{\alpha}{\alpha + \beta}(1 - \exp[-(\alpha + \beta)t])$$

$$\mu_{10} = 1 - \frac{\beta}{\alpha + \beta}(1 - \exp[-(\alpha + \beta)t])$$

III. Expected variance after time t

$$\sigma_{01}^2 = \mu_{01}(1 - \mu_{01})$$

$$\sigma_{10}^2 = \mu_{10}(1 - \mu_{10})$$

IV. Computation of a standard score

Assume that character X has changed from 0 to 1 over a branch of length 50 and that $\alpha = 0.01$ and $\beta = 0.01$

$$\mu_X = \frac{0.01}{0.01 + 0.01}(1 - \exp[-(0.01 + 0.01)50]) = 0.32$$

$$\sigma_X^2 = 0.32(1 - 0.32) = 0.22$$

$$\text{standardized score} = \frac{1 - 0.32}{\sqrt{0.22}}.$$

or beginning and ending in 1. By convention, however, we assign (arbitrarily) a positive sign to any branch ending in 1 and a negative sign to any branch ending in 0. This convention says, in effect, that a transition from 0 to 1 or a branch remaining in 1 are the same kind of evidence. For example, Y and X both changing from 0 to 1, or Y and X both remaining in state 1, are evidence that Y and X evolve together.

The set of scaled observations on X and Y obtained by applying eqns (4.5) to (4.8) to each of the branches of a phylogeny can be tested using a Pearson correlation coefficient, under the assumption that change or lack of change in each branch is an independent event, and that the underlying distribution of standardized Y and X values throughout a phylogeny is bivariate normal. However, because the model does not assume normality a non-parametric test of correlation might be used instead. Either test assumes that the characters are evolving independently in *each branch* of the tree, and that ancestral conditions are known independently of the species values. In practice, neither of these assumptions is likely to be met. Ancestral conditions will be reconstructed from the species' values. This introduces some dependency between the higher nodes and the tips. As a result, treating each branch of the tree as an independent data point probably overestimates the true number of degrees of freedom. For example, two sister species with the same character state will cause the reconstructed ancestral condition to be that state, whether it was in fact or not: the reconstructed ancestral condition, then, is not independent of the two species.

The extent of non-independence between the tips and the higher nodes depends upon the distribution of character states among extant species. An extreme example, that of all species having the same character state, fully determines all higher nodes. Further, the reason for a lack of character change in some branches may be shared among them. Without simulation studies of a variety of realistic situations, it is difficult to know how much bias will be introduced if all branches of the tree are counted in the test. A bifurcating phylogeny of n species will always have $2n-2$ branches, but probably closer to $n-1$ true degrees of freedom if the higher nodes are reconstructed from the tips. But consider a phylogeny with a large number of branches in which only a few show a character change. If the model developed here is used, and all branches are counted, a significant result is likely to emerge. One suggestion, then, is to use the data from all the branches but to assign only $n-1$ degrees of freedom to the statistic. A more conservative approach is to use all of the data to estimate likelihoods but only use in the statistical test the data from branches in which one or both variables change.

4.7 Estimating transition probabilities by maximum likelihood

It is necessary to have information on branch lengths, and on the probabilities of transitions in order to apply the model we have described. Branch length information is becoming increasingly available (see, for example, references in Springer and Krajewski 1989), and we assume for the moment that the limiting information will be on transition probabilities. It is possible to estimate transition probabilities, if one is willing to assume that the likelihood of a transition in a particular direction (e.g. 0 to 1) for a particular character is constant throughout the tree. We describe here a maximum likelihood approach for estimating the transition probabilities from real data.

We assume that a phylogeny is available along with branch lengths. The problem is to estimate the transition probabilities in each character in such a way as to maximize the likelihood of the particular distribution of characters given the phylogeny. We describe here how to find the transition probabilities for the character X. However, all of the equations for a second or other characters are the same. The overall likelihood of the distribution of character states in X for a particular tree and values of α and β is given by:

$$L(D) = \Pi(P_{01}(t_i))\Pi(1-P_{01}(t_j))\Pi(P_{10}(t_k))\Pi(1-P_{10}(t_l)). \qquad (4.9)$$

where the first product is found over the branches showing a transition from 0 to 1, the second product is over the branches showing 'a transition' from 0 to 0, and the third and fourth products are for transitions from 1 to 0 and 1 to 1, respectively. We seek the values of α and β that maximize $L(D)$ where "D" signifies that the likelihood is found for the data given the tree. Equation (4.9) can be expressed as a log-likelihood:

$$\log_e[L(D)] = \Sigma\log_e(P_{01}(t_i)) + \Sigma\log_e(1-P_{01}(t_j)) + \Sigma\log_e(P_{10}(t_k)) + \Sigma\log_e(1-P_{10}(t_l)). \qquad (4.10)$$

Analytical solutions to these equations will not be possible in most instances, so an iterative search procedure must be used to find the values of α and β that maximize equation (4.10).

In Box 4.4, we analyse the correlation between X and Y for the phylogeny in Figure 4.2 using maximum likelihood estimates obtained from an iterative solution of the log-likelihood equation (4.10).

Box 4.4. Standard score analysis of the data from Fig. 4.2 using maximum likelihood estimates of the transition probabilities

I. By maximum likelihood estimation

X: $\alpha = 0.00857$, $\beta = 0.00585$
Y: $\alpha = 0.010913$, $\beta = 0.014635$

II. Standard score analysis

Branch	X	Y	t	μ_X	μ_Y	S_X	S_Y	$\frac{X-\mu_X}{S_X}$	$\frac{Y-\mu_Y}{S_Y}$
$n_4 \rightarrow S1$ *	1→1	1→1	40	0.18	0.37	0.38	0.48	+0.47	+0.76
$n_4 \rightarrow S2$	1→1	1→0	40	0.18	0.37	0.38	0.48	+0.47	-1.31
$n_5 \rightarrow S3$ *	0→0	0→0	10	0.08	0.10	0.27	0.30	-0.29	-0.33
$n_5 \rightarrow S4$ *	0→0	0→0	10	0.08	0.10	0.27	0.30	-0.29	-0.33
$n_6 \rightarrow S5$ *	0→0	0→0	20	0.15	0.17	0.36	0.38	-0.42	-0.45
$n_6 \rightarrow S6$ *	0→0	0→0	20	0.15	0.17	0.36	0.38	-0.42	-0.45
$n_7 \rightarrow S7$ *	1→1	1→1	10	0.05	0.13	0.23	0.34	+0.24	+0.39
$n_7 \rightarrow S8$ *	1→1	1→1	10	0.05	0.13	0.23	0.34	+0.24	+0.39
$n_2 \rightarrow n_4$	1→1	0→1	50	0.21	0.31	0.41	0.46	+0.51	1.50
$n_2 \rightarrow n_5$	1→0	0→0	80	0.28	0.37	0.45	0.48	-1.61	-0.77
$n_3 \rightarrow n_6$	0→0	1→0	50	0.31	0.41	0.46	0.49	-0.66	-1.19
$n_3 \rightarrow n_7$	0→1	1→1	60	0.34	0.45	0.48	0.50	1.38	+0.90
$n_1 \rightarrow n_2$	0→1	0→0	70	0.38	0.36	0.48	0.48	1.28	-0.75
$n_1 \rightarrow n_3$	0→0	0→1	90	0.43	0.38	0.50	0.49	-0.87	1.27

* Branches along which neither character changes.
Note: signs of standardized scores are assigned according to the convention adopted in Section 4.6.1.

III. Spearman rank correlation test for correlated evolution of characters

For the 7 branches in which one or both variables change:-
$r_s = 0.25$, $P = 0.54$.
For all 14 branches:
$r_s = 0.31$, $P = 0.26$.

4.7.1 What if branch lengths of the phylogeny are not known?

Even when branch lengths are not known it should still be possible to improve upon merely ignoring the issue of transition probabilities. We can ask, for a given tree with unknown branch lengths, what values of α and β maximize eqn (4.10)? The transition probabilities and the branch lengths are inseparable as all branch lengths are assumed to be equal. We assume that their product is a constant and estimate it by maximum likelihood just as we did previously. Under these conditions, the maximum likelihood estimates of α and β will, at least, take into account the pattern of transitions in the phylogeny, if not the branch lengths along which they occur. Moreover, α and β take values in this case that yield reassuring results. When all branch lengths are assumed to be equal, the values of α and β will be such that the predicted character state for any branch beginning with 0 is simply the proportion of such branches that showed a change from 0 to 1, and the predicted character state for any branch beginning with 1 is just the proportion of such branches showing a transition from 1 to 0: for character X in Fig. 4.2, then the value 0.25 is predicted for all branches beginning in 0 and the value 0.167 is predicted for branches beginning in 1, assuming that all branches are of length 1.0.

4.8 Applying the statistical model to a real data set

Most species of the dog family Canidae can be classified as either carnivorous or omnivorous and also as either showing limited biparental or more extended communal care of the young; data are given in Gittleman (1983). Is there a relationship between diet and the evolution of communal care in canids as suggested by Gittleman (1985)? We can make use of a phylogenetic tree of the canids recently compiled by Wayne and O'Brien (1987) and Wayne *et al.* (1989) based on isozyme genetic distance data and supported in part by DNA hybridization (Fig. 4.7).

Fourteen carnivore species are included in the sample, providing a phylogenetic tree with 22 branches along which path lengths can be measured. The Spearman rank correlation between standardized scores for the two variables obtained by applying the procedures in Sections 4.6 and 4.7 is highly significant ($r_s = 0.55$, $P = 0.01$). However, perhaps more appropriately, using only the five branches in which one or both characters changes (Section 4.6), the correlation becomes non-significant ($r_s = 0.50$, $P = 0.32$).

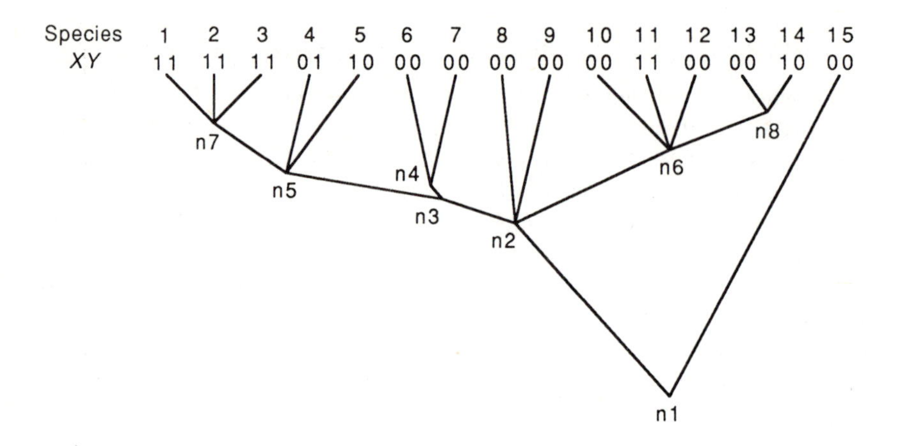

Fig. 4.7. Phylogenetic tree for fourteen canid species, with the black bear *Ursus americanus* used as an outgroup. Species 1 to 4 are: (1) grey wolf *Canis lupus*; (2) coyote *Canis latrans*; (3) cape hunting dog *Lycaon pictus*; (4) black-backed jackal *Canis mesomelas*; (5) bush dog *Speothos vanaticus*; (6) maned wolf *Chrysocyon brachyurus*; (7) crab-eating fox *Cerdocyon thous*; (8) grey fox *Urocyon cinereoargenteus*; (9) raccoon dog *Nyctereutes procyonoides*; (10) cape fox *Vulpes chama*; (11) red fox *Vulpes vulpes*; (12) fennec fox *Fennecus zerda*; (13) kit fox *Vulpes velox*; (14) arctic fox *Alopex lagopus*. Ancestral character states were calculated using Maddison *et al.*'s (1984) method (see Chapter 3, Box 3.1). Character X state 0 is omnivory, and state 1 is carnivory. Character Y state 0 is non-communal care, and state 1 is communal care. Approximate times before present (my) and ancestral character states (in parentheses) estimated according to the parsimony procedure described in Box 3.1, for each node are: n1 16.75 (00); n2 9.0 (00); n3 7.0 (00); n4 6.5 (00); n5 6.0 (11); n6 5.25 (00), n7 3.0 (11); n8 2.75 (00). (Phylogeny after Wayne *et al.* 1989).

4.9 Relationship of the statistical model to previous methods

4.9.1 The method of counting evolutionary events

Ridley's method for counting evolutionary events succeeds in deriving a set of independent data points, at least within the limitations of the methods used to reconstruct ancestral character states. However, each of the evolutionary events is given equal weight in the final chi-square or Fisher's exact test. But the model in the previous section shows how the likelihood of a character transition depends upon the length of the branch in which it occurs. What assumptions are implicit in assigning all branches equal weight?

Ridley's test can be understood in terms of the model presented in the previous section by noting that the chi-square statistic for a 2×2 contingency table is related to the correlation coefficient by the formula chi-square $= nr^2$ where n is the number of pairs of observations and r is the correlation coefficient between two dichotomous discrete variables (also known as the phi-coefficient). Thus, Ridley's test is simply a correlation of changes along the interior branches of a phylogeny, where all changes are assigned a 1 or a 0, and branches in which no change occurs are ignored.

The assumptions implicit in assigning equal weight to all transitions are made more clear by considering the example in Box 4.5. There we have calculated standardized scores after making all branch lengths in Fig. 4.2 equal to 10 and setting all four transition probabilities (α and β for X and Y) to a single value such that the standardized scores all take the values +1.0 or −1.0. Doing this yields a phi-coefficient (or equivalently a chi-square) for the overall test that is identical to that obtained by Ridley's test. This result will be true in general, not just for the example of Fig. 4.2. Ridley's test, then, in weighting all changes equally assumes implicitly that all branch lengths are the same, and that $\alpha = \beta$ for both characters (however α and β for X need not be equal to α and β for Y).

A further assumption of Ridley's model is that branches in which no change occurs are irrelevant to the hypothesis; that is, Ridley's method only counts branches in which one or both of the characters change. By ignoring branches in which no change occurs, Ridley's method is, in terms of the statistical model, assigning them a path length of zero (that is, the path does not exist).

4.9.2 The method of directional change

How does Maddison's test relate to the statistical model? The main feature of Maddison's test is that it considers all possible combinations of the ways that a certain pattern of transitions could evolve. As our worked example showed, this is equivalent to finding all possible redistributions on the tree of one of the characters, namely, the one taken as the dependent variable, while holding constant the placement on the tree of the independent variable transitions. Each of these redistributions is then counted once in arriving at the values for the numerator and denominator of Maddison's test.

But these different ways of assigning the transitions to the tree are not equally likely if branch lengths differ. We can estimate the likelihood for any possible redistribution of the characters in a tree in terms of the model in eqns (4.1) to (4.4). This statement would be equivalent to that given in eqn (4.9) and would be calculated for the transitions in the dependent variable. In terms of Maddison's model, a different probability statement can be calculated from eqn (4.9) for each of the possible redistributions of

Box 4.5. Comparison of Ridley's method of counting evolutionary events with the general statistical model, assuming equal branch lengths and transition probabilities

The data from Fig. 4.2 are used, but path lengths are assumed to be equal.

I. Method of counting evolutionary events

A tally of end-states in branches showing changes in *X* or *Y* produces the 2 x 2 contingency table:

		X 0	*X* 1
Y	0	2	2
	1	1	2

$\chi_1^2 = 0.19$ $\quad \phi = \sqrt{\frac{0.19}{7}} = 0.17$ $\quad P = 0.66$

II. Statistical model

Assume all branch lengths = 10

Set $\alpha = \beta = 0.50$

From eqns (4.5) and (4.7) or Box 4.3:

$$\mu_X = \frac{0.5}{(0.5 + 0.5)}(1 - \exp[-(0.5 + 0.5)10]) = 0.50; \mu_Y = 0.5.$$

From eqns (4.6) and (4.8):

$$\sigma_X = \sigma_Y = \sqrt{0.5^2} = 0.5.$$

This general result is true for *any* common path length. For example, if all branch lengths are 50 and the transition probabilities are 0.10, the means and standard deviations are still 0.50. In this example, the standardized scores are arbitrarily given the same sign if the end-states match: +1 for end-states = 1, -1 for end-states = 0 (see end of Section 4.6.1). This is equivalent to Ridley's method.

Branch	End-State		Standardized score	
	X	*Y*	*X*	*Y*
n4→*1* *	1	1	1	1
n4→*2*	1	0	1	-1
n5→*4* *	0	0	-1	-1
n5→*3* *	0	0	-1	-1
n2→n4	1	1	1	1
n3→n6	0	0	-1	-1
n7→*7* *	1	1	1	1
n7→*8* *	1	1	1	1
n6→*5* *	0	0	-1	-1
n1→n2	1	0	1	-1
n1→n3	0	1	-1	1
n3→n7	1	1	1	1
n2→n5	0	0	-1	-1
n6→*6* *	0	0	-1	-1

When branches with no change (marked *) are ignored, a 2 x 2 contingency table comparing the number of branches for which the standardized *X*s and *Y*s take the same or different values (+1 or -1) produces the same result as the previous method:

		Standardized *X*	
		-1	1
Standardized *Y*	-1	2	2
	1	1	2

Association statistics are, of course, also identical.

The analysis derived from Ridley's model is seen to be equivalent to the general model if all branch lengths are assumed to be equal, and $\alpha = \beta$ for each character. These circumstances will yield standardized scores of ± 1. Ridley's method does not count branches along which there is no change of character state.

the character transitions on the tree. These different likelihood statements will not all take the same value, because, depending upon branch lengths, a transition may be more or less likely in one part of the tree than in another. The sum of the probabilities of all possible redistributions would conform to Maddison's quantity b. But now this quantity would not assign each redistribution an equal weight, but rather a weight based on the probability of that particular redistribution. A subset of the possible redistributions of the character on the tree will conform to the quantity a in Maddison's test. The test statistic would then be a weighted version of Maddison's where each of the possible redistributions pertaining to Maddison's quantities a and b would be weighted by its probability of occurrence:

$$P(Y \mid X) = \frac{\Sigma\,(y_a)}{\Sigma\,(y_b)} \tag{4.11}$$

where $P(Y|X)$ is the probability of the character transitions in Y (the dependent variable) arising by chance given the distribution of the X variable transitions, y_a are the weighted redistributions according to Maddison's quantity a, and y_b are all of the possible weighted redistributions.

We have calculated the different likelihoods for the nine possible redistributions of character states for the phylogeny of Fig. 4.4, according to the branch lengths shown in Box 4.6, and using the transition probability derived from the maximum likelihood solution. In this instance, only α is estimated since there are no back transitions in the tree, and thus $\beta = 0.0$. Not surprisingly, some of the redistributions are substantially less likely than others. For example, the redistribution corresponding to the two transitions occurring in the shortest branches of the tree is about 1/60 as likely as the tree with the two transitions in the longest branches. The ratio test for the observed tree (Fig. 4.4) is 0.076. This compares with 1/9 or 0.111 for Maddison's test. The statistical model gives a slightly smaller probability in this instance by recognizing that one of the transitions occurred in a very short path segment.

Like Ridley's method, the implicit assumptions of Maddison's test in assigning all possible redistributions of characters equal weight are that all branch lengths are equal and both transition probabilities (Maddison's test only studies transitions in the 'dependent variable') are the same. To see why, consider that each of the possible redistributions must be given the same probability by eqn (4.9). That is, all of the ys in eqn (4.11) must take the same value. Making all the branch lengths equal in the likelihood statement of eqn (4.9) and choosing α and β appropriately makes all the transitions have a probability of 0.5. Then, each redistribution has the

Box 4.6. The ratio test for the phylogeny shown in Figure 4.4

I. Assume unequal branch lengths as shown in the Figure below:

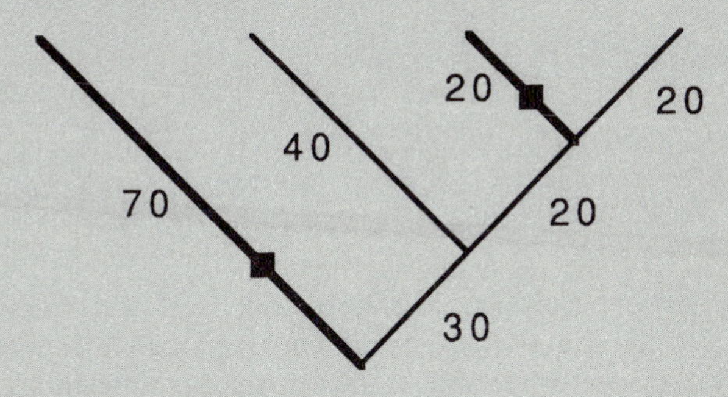

The maximum likelihood estimate for α in this tree is 0.0134. Using this value, we can find the likelihood for each of the nine possible arrangements in Box 4.1.

Arrangement (from Box 4.1)	Likelihood $y = \prod(1 - e^{-\alpha t})\prod(e^{-\alpha t})$
1	0.39 * 0.67 * 0.59 * 0.77 * 0.24 * 0.24 = 0.0065
2	0.39 * 0.67 * 0.42 * 0.77 * 0.24 * 0.77 = 0.0149
3	0.39 * 0.67 * 0.42 * 0.77 * 0.77 * 0.24 = 0.0149
4	0.39 * 0.67 * 0.42 * 0.24 * 0.77 * 0.77 = 0.0149
5	0.61 * 0.67 * 0.59 * 0.77 * 0.77 * 0.24 = 0.0328
6	0.61 * 0.67 * 0.59 * 0.77 * 0.24 * 0.77 = 0.0328
7	0.61 * 0.67 * 0.59 * 0.24 * 1.00 * 1.00 = 0.0560
8	0.61 * 0.67 * 0.42 * 0.77 * 0.77 * 0.77 = 0.0757
9	0.61 * 0.33 * 1.00 * 1.00 * 1.00 * 1.00 = 0.2016

$$P(Y|X) = \frac{\Sigma(y_a)}{\Sigma(y_b)}$$

For example, $P(Y|X)$ for arrangement 6 $= \frac{0.0328}{0.4501} = 0.073$

II. Assume, instead, that all branch lengths are equal. If all branch lengths = 70 and $\alpha = 0.0099$, it follows that the probability of each path, with or without a transition, is 0.50. Then:

$$y = (0.50)^6 \text{ for each redistribution}$$

$$P(Y|X) = \frac{\Sigma(y_a)}{\Sigma(y_b)} = \frac{(0.50)^6}{9(0.50)^6} = \frac{1}{9} \text{ for any redistribution}$$

Thus, Maddison's test makes all arrangements equally likely by setting path lengths equal and choosing transition probabilities to be the same.

same probability of occurrence, and the overall likelihood of any particular outcome is just the sum of the (now all equal) probabilities that make up the quantity a divided by the sum of the (all equal) probabilities that make up the quantity b, or a/b.

4.10 Discussion

This chapter has described Ridley's method for assessing correlated evolutionary change in dichotomous variables, and Maddison's method for detecting the direction of evolutionary change. We showed how both of these methods could also be thought of as special cases of a more general statistical model that takes into account information on branch lengths. The general model can be applied to any data set by using the maximum likelihood procedures for estimating the model's parameters.

In the hypothetical examples that we used to introduce the statistical model, we assumed that the phylogeny and the ancestral states were known without error. This luxury of course will seldom be obtainable with real data sets, and so it is important to recognize some of the limitations of the tests that we have discussed. One difficulty concerns the assumption that each of the branches of the tree can be used as an independent data point in testing for the correlation of changes between Y and X. In a bifurcating tree with n tips, there will always be $2n$–1 apparent degrees of freedom for analyses. However, for two reasons there may be fewer actual degrees of freedom. The first is that if, for example, an ancestral node is in state 0, then each daughter node that radiates from it is constrained either to change to 1 or not at all: they cannot change from 1 to 0. Second, in most instances the ancestral nodes will not be known and will have to be estimated from the species values. This introduces a dependence between the ancestral and descendant states that is not easily quantified, except in extreme cases. For example, if all species have the same value of the character, then all higher nodes will be reconstructed to have that character (see also discussion at end of Section 4.6.1).

Another difficulty is the assumption that the transition probabilities αdt and βdt are constant throughout the tree. If these two parameters vary considerably, then the scalings of X and Y will be incorrect. There is good reason to expect that αdt and βdt will vary (see Diamond and May 1977 for an example from island biogeography). However, this is not a limitation of the formal models (indeed the problem also plagues Ridley's and Maddison's methods), but of our understanding of which taxa should and should not be included in the same analysis. A related problem, perhaps, is that ancestral character states frequently will have been found according to a parsimony rule, but the statistical model assumes a Poisson process underlying the changes of character states. The effect this might have on the performance of the method has not been investigated.

Computer simulation studies, such as we report in Chapter 5, are needed to determine how seriously these issues may affect the statistical tests. Preliminary evidence concerning a problem of non-independence of branches with a method for continuous variables suggest that the effects may not be serious (see Section 5.10). Lacking evidence from simulation studies, there seem to be two courses of action, one more conservative than the other. The first involves reducing the number of degrees of freedom for statistical tests. Felsenstein (1985*a*) shows that n–1 independent comparisons between species and higher nodes can be derived from a bifurcating phylogeny with n species. Using this number would approximately reduce by half the number of degrees of freedom compared to counting all branches as independent points. A more conservative approach is to use in the final analyses only those branches in which one or the other character changes.

Another class of solutions would avoid conditioning the comparative test on any one particular set of higher nodes[10]. The idea is to estimate the four transition probabilities (X and Y changing from 0 to 1 and 1 to 0) independently of each other from the counts of the numbers of each kind of transition throughout the phylogeny. For any set of transition probabilities it is possible to calculate the likelihood of all possible reconstructions of ancestral character states. This likelihood can be compared with that derived from the set of transition probabilities that maximize the likelihood of finding the extant character states. Much work remains to be done in this area.

The examples that we have used to illustrate the model have all used bifurcating phylogenies. Even if some branches of the tree are not resolved

---

[10] We thank Joe Felsenstein for this idea.

to this level, however, it is still possible to calculate standard scores. A multifurcating node can be thought of as representing two or more bifurcating nodes joined (implicitly) by a path or paths of zero length. The issue, then, is whether there have been any transitions along these paths. Probably the safest way to deal with a multifurcating node is to calculate only one standard score for each end-state represented among the sub-nodes (i.e. up to a maximum of two; see also Maddison 1989).

We have chosen a particular evolutionary model to describe the evolution of two dichotomous characters. More realistic alternatives may be possible. For example, rather than conceptualizing transitions between characters states as discrete jumps, there may be an underlying quantitative dimension whose phenotypic expression is two character states. Whether an individual is in one or the other state would depend upon the effects of many genes. Beyond a certain threshold along the dimension, an individual has one of the states, below the threshold it has the other (see discussion in Felsenstein 1988). A feature of the threshold model would be that not all individuals in the same state have the same probability of changing to the other: those closer to the threshold are more likely to change. Although the mathematics of this method are not yet worked out, it seems a promising approach.

Statistical and mathematical models of evolution are often criticized as biologically implausible or unrealistic. However, even techniques that apparently are not based on an evolutionary model may in fact just be special cases of *implicit* evolutionary models. These special cases may be even less realistic than the models. Thus, even if models are biologically unrealistic, they serve important functions. One is to make explicit the consequences of the assumptions made implicitly when using existing methods. For example, different redistributions of character changes on a phylogenetic tree varied 12-fold in their likelihood of occurrence according to the statistical model, and yet all were given the same weight by Maddison's method. An equally important function of models is that they force us to think clearly about the sorts of processes that are thought to give rise to the phenomena we are attempting to understand. This will, no doubt, lead to more complicated models that attempt to incorporate a greater number of the factors thought to be giving rise to the observed phenomena. Nevertheless, it is important to bear in mind that, in practice, results will depend partly on the true phenomena under study, and partly on the particular assumptions of the methods used. Where the conclusions depend on the model used, this should be acknowledged explicitly, and some justification should be given for choosing one model over another.

4.11 Summary

This chapter develops a general model for the comparative analysis of discrete data. The model is designed to be used in conjunction with a phylogeny for which branch lengths, and the probabilities of character change, are known. A maximum likelihood estimation procedure is described for estimating the latter, even when branch lengths are not known. Thus, the model can be applied to existing data sets. Existing methods for the analysis of categorical data can be derived as special cases of this model, by making particular assumptions about branch lengths and probabilities of character change.

5

Comparative analysis of continuous variables

'Independent evolution may be the ideal criterion for the comparative method' (Ridley 1983*a*, p. 18)

5.1 Introduction

Many of the questions of interest to comparative biologists involve comparing the values of characters that vary continuously, rather than discretely, among a number of species. Measures of morphology, physiology, life histories, and behaviour, such as walking, flying, or swimming, produce quantitative values. Comparative studies of characters that vary continuously have progressed from relying nearly exclusively on cross-species correlational analyses that ignore the historical relationships among species, to sophisticated techniques that incorporate information about phylogenetic relationships into the comparative test. We review briefly each of the major approaches in this chapter. Those comparative methods that compute sets of statistically independent comparisons, either across contemporary taxa, or between ancestral and descendant nodes, emerge as the best techniques currently available. This conclusion is supported by the results of computer simulation studies.

As with tests for discrete characters, tests for continuously varying characters are confronted with the problem that species form a nested hierarchy of phylogenetic relationships. For the reasons described in Chapter 2, closely related species are typically more similar than are more distantly related species. This means that species cannot be treated as independent units of information in statistical tests. The various comparative techniques that we review in this chapter can be distinguished by the procedures that they use to manage the effects of similarity associated with phylogenetic relationships. Each of the methods is based on a set of, often implicit, assumptions that comprise a null hypothesis of evolutionary change, and on a set of statistical techniques that apply those assumptions to real data. The statistical techniques are designed to produce data points that can be treated as independent for statistical analysis.

Should we even bother with a separate class of methods for continuous variables? In a mathematical sense, the distinction between models for discrete variables and models for continuous variables hides a deeper equivalence between the two cases. Continuous variables, the topic of this chapter, are merely discrete variables in which we allow the width of the discrete units to become vanishingly small. Assume that change in a continuous metrical character proceeds by a series of discrete steps. The 'distance' moved each time is a fixed amount either 'forward' one unit (+1) or 'backward' one unit (−1) along the scale. Forward and backward movements are equally likely. Let there be a single step in each of a number of successive epochs of time, where a single epoch is denoted by τ. The position along the scale after n such epochs is the sum of all the preceding steps. These circumstances give rise to what is known as a random walk. Some random walks will have a preponderance of +1s, others a preponderance of –1s, others a more even number of pluses and minuses. Given that one of two possible outcomes occurs at each interval of time, there will be 2^n possible outcomes of the random walk after n steps.

Statistical theory informs us about the expected distribution of outcomes of such a process when we allow the unit of time τ to become vanishingly small, and thereby allow n to become large. It can be deduced from the Central Limit Theorem that, in the limit as τ goes to zero, the outcomes of the random walk after some unit of time t will be normally distributed with an expected value of zero, and variance of $\sigma^2 t$, where σ^2 is a positive constant. In continuous time, the discrete process is transformed into one of Brownian motion where the variance of change accumulates in direct proportion to the amount of time the process has been allowed to go. We make use of this result in the next section to illustrate various properties associated with phylogenies.

We begin this chapter with a discussion of the statistical problems that arise when data obtained from phylogenies are used in conventional statistical procedures. These are the same problems that arose in connection with methods for discrete characters, but now we discuss them using statistical models that are more convenient for continuously varying characters. We then show how each of the more recently developed techniques confronts those statistical problems. Although our account is somewhat historical, it also develops as a logical progression of ever more acceptable procedures. Our final sections present the most recently developed techniques, and then summarize selected results from computer simulations which serve to underline one of the main messages of this book: we make assumptions about the way evolution proceeds whenever we choose a comparative test.

5.2 Testing hypotheses on continuous variables

The simplest way to test for a relationship between two continuous variables is to treat species as independent data points and apply standard statistical techniques to characterize their relationship. This was also true for testing the relationship between two discrete variables. However, as we pointed out in Chapters 2 and 4, such an approach is bound to be flawed statistically because species are part of a hierarchical phylogeny. This fact virtually guarantees that each species will not have independently evolved the suite of traits that defines its phenotype, thus posing a critical problem for statistical methods which assume that the data points are independent.

To see why the preceding statement is true, it is necessary to think about the various ways that, for any given set of contemporary species, evolution could have arrived at the present. Consider the phylogeny of the eight contemporary species in Fig. 5.1. This figure is similar to Fig. 4.1, except now the species values represent two continuous characters.

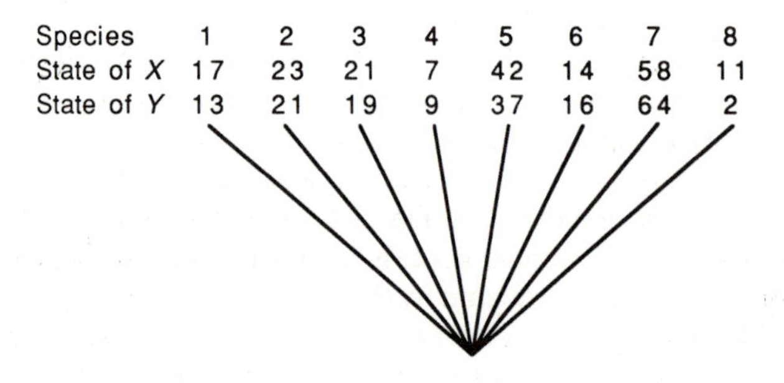

Fig. 5.1. Eight species simultaneously evolving from a common ancestor. Values of two continuously varying characters (X and Y) are given for each species. Figure 4.1 uses the same example with discrete characters.

For the purpose of discussion, we assume that characters evolve independently according to a Brownian motion process, following Edwards and Cavalli-Sforza (1964) and Felsenstein (e.g. 1981*b*, 1985*a*, 1988). This statistical model is appropriate for describing the random wanderings of a variable along a continuous dimension (see Section 5.1). Figure 5.2 plots four random walk sequences in which the value of the random walk at any point is the sum of all of the changes before it. The four sequences can be thought of as characterizing the evolutionary changes through time in any four species from Fig. 5.1. The displacements along the horizontal axis represent the value of a character, and the vertical axis is time.

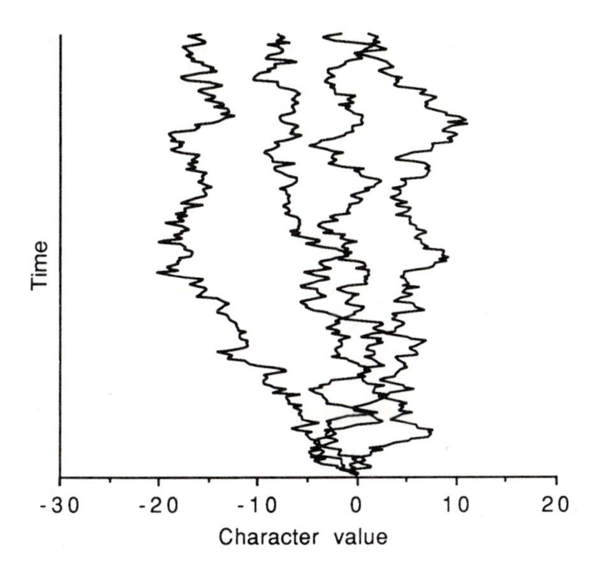

Fig. 5.2. Character change among four independently evolving lineages following random walks after splitting from a common ancestor at time zero. The value of each random walk at any point is the sum of all previous changes for that lineage. See also Felsenstein (1988).

Alternatively, the sequences in Fig. 5.2 can be thought of as iterations of the same random evolutionary process for a single species. This interpretation makes explicit what is meant by the expected variance of evolutionary change. If we were to re-run the random walk many times, the variance of the end-points would be an estimate of the expected variance of evolutionary change for that particular branch length. More generally, if the variance of a single step of the random walk is σ^2, then the expected variance of end-states after time period t will be $t\sigma^2$.

Thinking about the evolution of two or more species in terms of random walks can be used to illustrate why phylogenies might introduce correlations among species. If the phylogeny of species is like that in Fig. 5.1, then the random walks illustrated in Fig. 5.2 make the point that the evolutionary changes in the species since their common starting point have been independent. Thus, given the Brownian motion model, and the phylogeny of Fig. 5.1, we can treat species as independent points for statistical purposes. Furthermore, because the statistical properties of the random walk are known, we can say something about the expected change in each branch, as well as the expected variance of change in each branch.

But now consider that the eight species have the phylogeny shown in Fig. 5.3. This is, again, similar to a figure (Fig. 4.2) used in Chapter 4, except now the characters are continuous.

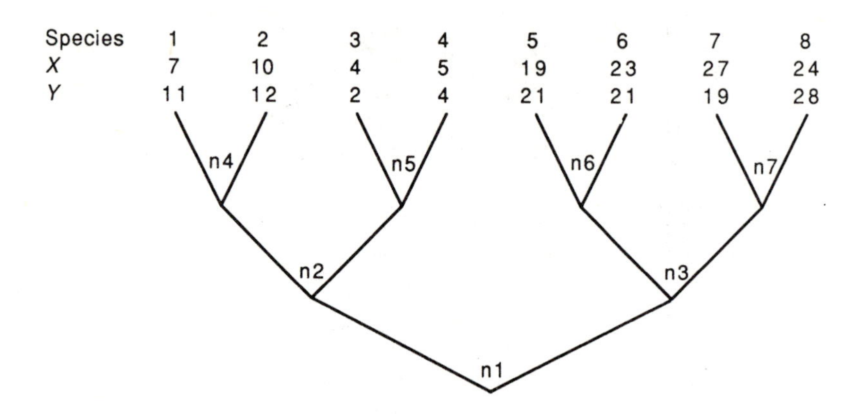

Fig. 5.3. A dichotomously branching phylogenetic tree showing the evolutionary history of eight species. Values of two continously varying characters (X and Y) are given for each species.

It will often be observed in such a phylogeny that the species in each pair of tips will be more similar to each other than to the other species. The Brownian motion model along with the phylogenetic structure can be used to illustrate this phylogenetic similarity. Figure 5.4 displays the results of a random walk that could correspond to species 1 through 4 from Fig. 5.3. At the beginning of the sequence there is but one path, indicating the common ancestry of all four species. At a later time the bifurcation corresponding to node n2 occurs. Later yet, nodes n4 and n5 occur. At the end of the sequence species 1 and 2 are closer to each other than to species 3 and 4. Their shared history introduces a correlation between them, even though the evolutionary changes along all branches of the tree have been independent. The sobering message from Fig. 5.4 is that phylogenetic similarity can arise from a completely random process. Imagine the similarity that arises when we allow for the processes described in Chapter 2!

Similarity associated with phylogeny causes statistical problems. Most statistical methods assume that the data points can be thought of as (1) having been sampled independently from (2) a normal distribution with some mean and variance. Phylogenetic similarity, however, by introducing a correlation among characters, invalidates the first assumption. The second assumption is slightly more complicated. Anticipating issues to be discussed later in this chapter, there are two main ways that points sampled from a phylogeny may have different expected variances.

First, we have assumed that the Brownian motion process proceeds at the same rate everywhere in the phylogeny. This guarantees that over any given amount of time, all lineages with a common starting point will have

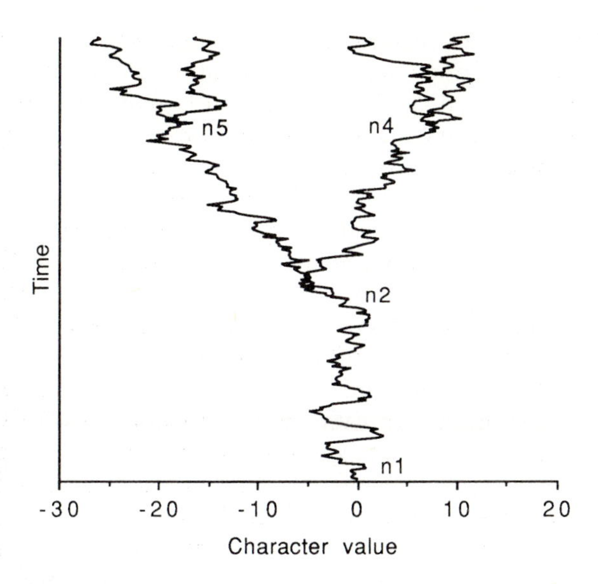

Fig. 5.4. Character change among four lineages following random walks after splitting from a common ancestor at time zero. The value of each random walk at any point is the sum of all previous changes for that lineage. Unlike the lineages in Fig. 5.2, the ones represented here have not evolved independently but have shared varying amounts of common ancestry since time 0, forming part of the phylogenetic tree shown in Fig. 5.3 There is a lineage split at n2, and two subsequent splits at n4 and n5. See also Felsenstein (1988).

the same expected variance of change. However, if we relax the biologically unrealistic assumption of equal rates of change (Simpson 1945), different expected variances of change occur in different lineages, thereby invalidating the assumption of equality of variances. Second, inequality of variances may also arise even if σ^2 is fixed if we compare two or more values derived from the phylogeny that do not have a common starting point, and which therefore may not have been evolving for the same length of time.

We have developed these points in detail because they illustrate all of the essential statistical and evolutionary considerations that must go into developing a test of a comparative relationship. We have statistical problems because phylogenetic relationships cause a lack of independence among the data points, and because unequal rates of change or differing time periods of change in different lineages may introduce inequality of variances. Fundamentally the same problems were encountered in Chapter 4. Either of these two problems—lack of independence or heterogeneity of variances—renders many standard statistical tests invalid. There are statistical techniques for managing the effects of non-independence and unequal variances (Generalized Least Squares, see Draper and Smith 1981), but they depend upon being able to specify the nature of the correlation among species, and the expected variances of their characters. This is where the evolutionary considerations become central: our beliefs about the extent of the correlation among species and the extent to which variances are unequal depend upon our assumptions about how evolution proceeds. Does evolution move along at a constant rate in all branches or does it have different rates in different branches? Alternatively, is evolutionary change punctuational such that branch length is less important? Or, is it such a jumble of these two processes (and others) that it is futile to assume one or the other? The answers that we give to these questions are crucial and, as with similar questions in Chapter 4, they define the various different methods of comparative analysis.

There cannot be 'solutions' to the problems posed by comparative data, then, only approximations to solutions based upon our current understanding of evolution. Some approximations will be demonstrably better than others for particular situations but, in many instances, the validity of a technique will be unknown. This is an important point because it means that the choice of a particular approach for analysing comparative data will often depend less upon knowledge that one technique is superior to another, than on a set of beliefs about the workings of evolution for a particular set of species and variables.

In the sections that follow, we describe different methods that have been used to analyse comparative data on continuous variables. Each method can be characterized by the statistical techniques that it uses to produce (at least in theory) data points for comparative analysis that are independent, and that all have the same expected variance. Some of the methods attempt to manage the effects of phylogenetic similarity by estimating the extent of non-independence and heterogeneity of variance in the data on the basis of the phylogenetic tree. Other methods create independent data points by discarding the variation that is thought to reflect phylogenetic similarity. An additional class of methods based on equally plausible models can test the comparative relationship without discarding any of the information. This class of methods attempts to define a set of mutually independent comparisons calculated from the phylogeny. Comparisons can be either between species or between higher nodes, or between the beginning and end-states along a branch, in which case the direction of phylogenetic change is of interest (directional versus non-directional comparisons are discussed in Chapter 1). In either case, the comparisons are then scaled according to empirical or model-based rules that attempt to equalize their variances. We argue that these techniques are the best currently available for conducting comparative tests.

For the rest of this chapter, we shall make repeated reference to Figs 5.1–5.4 which are drawn, following convention, with extant species at the top. Unfortunately, there is another convention that we must follow, and that is to talk of 'higher' versus 'lower' taxa, and 'higher' versus 'lower' nodes. As far as phylogenetic trees are concerned, typified by Figs 5.1 and 5.3, higher taxa and nodes are usually drawn at the bottom of the tree and lower taxa and nodes are at the top! For example, a species is a lower taxon than a family. It seems unnatural to break either of these conventions, and appropriate care must thus be taken when translating between the text and the figures.

5.2.1 A cautionary note

Unless otherwise stated, we assume in our discussions of the various methods that the phylogenies are known without error. This, of course, will seldom be true, but it is an important assumption, given the uncertainties about whether a reconstructed phylogeny is the true phylogeny, even when the best methods are used (see Chapter 3). Some work has been done on the issue of placing confidence limits on estimates of phylogenies (see Felsenstein 1985*c*, and references therein). Little is known about how sensitive the conclusions of a particular study will be to the tree that is used. Until more work is done in this area, comparative biologists should be aware that their results may depend upon a particular reconstruction of a phylogeny. If several phylogenies are equally likely (or parsimonious, or compatible), then perhaps the analyses should be done using them all (see Björklund in Harvey 1991 for an example). If the conclusions vary widely, caution should be exercised in their interpretation.

5.3 Species analyses

Species values have been used as the units for statistical analysis in the vast majority of comparative studies that have analysed relationships between continuously varying characters[11]. We have stated that species cannot be assumed to be independent and that they may have different expected variances. So, what evolutionary assumptions are implicit in using species as the units of analysis? That is, what models of evolution would give rise to species being independent and having the same variances?

As was true for discrete characters (following Fig. 4.1), the model implicit in Fig. 5.1 provides one answer. Under this model of evolution, to use species as the units of analysis means that the investigator must be

[11] Examples from a variety of taxa and fields of biological enquiry include: Quiring (1941); Newell (1949); Hutchinson and MacArthur (1959); Southwood (1961); McNab (1963); Schoener (1968); Millar (1977); Clutton-Brock *et al.* (1977); Wooton (1987); Burt and Bell (1987); Clutton-Brock (1989).

willing to defend the belief that the phylogeny of the species can be represented by a simultaneous radiation of all of the species from a single common ancestor (see also Felsenstein 1985*a*). That is, the investigator must assume that there are no shared branches in the phylogeny. It must also be assumed that rates of evolutionary change are the same in each branch, an assumption guaranteeing that each datum has the same expected variance. These assumptions, if met, would mean that the different species could be treated as random samples from some common underlying distribution of possible outcomes. However, true simultaneous radiations may be rare in nature, and do not characterize the phylogenies of interest to most biologists.

Thus, to use species as independent data points in a comparative analysis requires that one ignores phylogenetic relationships. This should be anathema for anyone who believes in evolution. Nevertheless, species are used this way in comparative analyses, and so it is prudent to be aware of the consequences. The major problem is that the confidence limits on statistics are sensitive to the number of degrees of freedom declared in the analysis. Because species are typically not independent, confidence limits will be spuriously narrow. This can lead to rejection of a null hypothesis on false grounds. Allometric studies that employ species provide good examples.

Figure 5.5 shows a logarithmically scaled plot of home range size against body weight for 72 primate species. The slope of the model 1 regression line describing the best fit relationship between home range size and body weight is 1.26, with 95 per cent confidence limits from 0.95 to 1.58. A slope of 0.75, which might have been expected on energetic grounds (Kleiber 1961), lies outside these confidence limits. However, it is not valid to reject

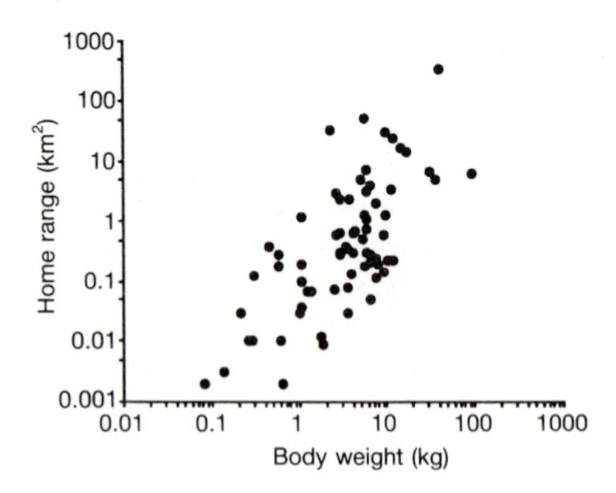

Fig. 5.5. Home range size increases with body weight across a sample of 72 primate species. When both axes are scaled logarithmically, the relationship is approximately linear with a slope of about 1.26. (Data from Harvey and Clutton-Brock 1981, and subsequent sources).

0.75 for this reason alone. The sensitivity of the analysis to the number of degrees of freedom is illustrated by the fact that if the sample size is reduced to 30, which is approximately the number of genera in the sample, the confidence limits widen to embrace 0.75.

5.4 Analysis of higher nodes

If species cannot be considered independent then perhaps some higher node can be? Crook (1965) suggested using genera and family means rather than species, but it was not until more than a decade later that an explicit statistical criterion was offered for this practice. Clutton-Brock and Harvey (1977), Harvey and Mace (1982), and Harvey and Clutton-Brock (1985) developed the use of the nested analysis of variance (Sokal and Rohlf 1969) to describe how the total variation among species in a continuous character is distributed among the taxonomic levels. The distribution of variance by taxonomic level is, in turn, used to identify which taxonomic level to use as the unit of analysis.

The nested ANOVA partitions the total variation among species into components representing each of the nested levels in a taxonomy:

$$\sigma^2_{tot} = \sigma^2_{s(g)} + \sigma^2_{g(f)} + \sigma^2_{f(o)} + \sigma^2_{o(c)}. \tag{5.1}$$

The term on the left is the total variance among species on the trait of interest. This total variation is then partitioned into, to adopt a simple taxonomy, a component representing the variation of species within their genera, the variation of genera within their families, families within orders, and finally orders within the class. At each level the mean of the values from the level immediately below is used.

If both sides of equation (5.1) are divided by $\sigma_{tot}{}^2$ and multiplied by 100, then the left hand side must be 100, and the terms on the right hand side become the percentages of variance found at each taxonomic level:

$$(\sigma^2_{tot}/\sigma^2_{tot}) \times 100 = [(\sigma^2_{s(g)} + \sigma^2_{g(f)} + \sigma^2_{f(o)} + \sigma^2_{o(c)})/\sigma^2_{tot}] \times 100. \tag{5.2}$$

These percentages of variance then can be compared among variables with different total variances. Another way to use the terms is to express them as cumulative proportions of variance moving from the highest level to the lowest level. Thus, if species is the lowest level of analysis then, by the time the $\sigma^2_{s(g)}$ term is added, 100 per cent of the variance will be accounted for. These cumulative percentages of variance have a precise statistical interpretation as *intra-class correlations*. The interpretation of an intra-class correlation is the correlation expected between any two data points selected at random from the same group. For example consider that species are the lowest taxonomic level represented and 75 per cent of the variance is accounted for by the combination of orders within the class and families within the orders. Then, consider sampling repeated pairs of species where the first member of each pair is chosen at random from the data set and the second is another randomly chosen species from the same family as the first. The correlation among the pairs will be 0.75. The intra-class correlation coefficient might be viewed as a measure of the power of one species for predicting the value of another species in the same family.

For many continuous variables, nested ANOVA reveals that most of the variation in the trait occurs among orders nested within the class, and among families within orders (although, in principle, there is no constraint on how the variation is partitioned). Body weight in mammals provides a good example. Different orders tend to vary a great deal—compare Proboscidea (elephants) with Chiroptera (bats)—whereas species within genera tend to have relatively similar body weights. Table 5.1 displays the results of nested analyses of variance on several size and life history variables in mammals.

Table 5.1 Taxonomic distribution of life history variance among placental mammals. Tabulated values are percentages of total variance accounted for at successive taxonomic levels estimated from a nested ANOVA on logarithmically transformed species averages. (After Read and Harvey 1989).

Among: Within:	species genera	genera families	families orders	orders class
Variance component:*	$\sigma^2_{s(g)}$	$\sigma^2_{g(f)}$	$\sigma^2_{f(o)}$	$\sigma^2_{o(c)}$
Adult weight	3	7	21	69
Neonatal weight	3	5	27	65
Gestation length	2	6	21	71
Age at weaning	8	11	19	62
Maximum reproductive life	10	10	12	68
Annual fecundity	5	7	14	74
Annual biomass production	6	8	18	68

*Following eqn (5.2), each variance component in the table has been multiplied by 100 and divided by σ^2_{tot}.

Harvey and Mace (1982) suggested that patterns similar to those in Table 5.1 might be interpreted to indicate that species within genera, and genera within families provide additional but probably not independent data points in an analysis. That is, any one species' value will tend to be a good predictor of the other species in that genus, and the mean for a genus will tend to predict the other genera in a family relatively well. An alternative interpretation of cumulative percentages of variance (obtainable by adding the successive components of Table 5.1) as intra-class correlations supports this interpretation. Two randomly chosen individuals from the same family will typically correlate highly. Moving down a level to genera does not increase the correlation much: genera do not add substantially new information.

The nested ANOVA, then, provides a suggestion of the taxonomic level that should be used for analysis. Families or orders would be chosen as the units of analysis for the variables in Table 5.1, because around the family level there is a precipitous decline in the percentage of variance that a taxonomic level accounts for (statistical tests are available: Sokal and Rohlf 1981). However, any taxonomic level could, in principle, be chosen depending upon the distribution of variance in the characters being studied. Choosing a higher node greatly reduces the number of degrees of freedom in analyses. This accords with the belief that the additional 'degrees of freedom' obtained from lower taxonomic levels are not really very free at all. Thus, degrees of freedom and some variability are given up in return for what is hoped are increasingly independent data points.

The higher nodes method represents an early attempt to give a statistical justification for not treating species (or genera or even some higher taxa) as independent points. In addition, it has often been used to suggest a taxonomic level at which independence can be more or less assumed[12].

Harvey and Zammuto (1985) used the higher nodes method to investigate life history variation in mammals. These authors were interested in the idea that mortality patterns should be strongly correlated with the age at which individuals reach maturity, independently of adult body weight. Others had argued that this need not be so (e.g. Western and Ssemakula 1982). Across species, we may test the prediction that variation in mortality rates should correlate with the age at which different species reach maturity when body size is held constant: high mortality rates must be associated with early ages at maturity lest individuals die before successfully reproducing.

Harvey and Zammuto used Millar and Zammuto's (1983) data on age at maturity and life expectancy in 29 mammal species. Life expectation at

[12] The method has been used to investigate variation in life history, morphology, metabolism, sleep, and other behaviours including patterns of habitat utilization, particularly in birds and mammals, (e.g. Harvey and Clutton-Brock 1985; Gittleman 1986; Harvey *et al*. 1987; Elgar and Harvey 1987; Elgar *et al*. 1988; Bennett and Harvey 1985*a*, *b*, 1987; Read and Harvey 1989; Sherry *et al*. 1989; Promislow and Harvey 1990).

birth measured from natural populations of mammals living in approximately constant age-structured populations was taken to be an inverse measure of adult mortality rate. Harvey and Zammuto chose to analyse genera means instead of species because of concerns that the individual species did not represent independent points. When all variables were logarithmically transformed, life expectancy and age at maturity were both positively correlated with body weight (r = 0.87, 0.89, respectively, n = 25, P <0.001). However, the correlation between the two life history variables remained significant after controlling for the effects of body weight (r = 0.89, n = 25, P <0.001). This correlation may arise, however, because life expectancy is partly a function of age at maturity (demographic reality dictates that some individuals must survive to breed or the species would be extinct). So, Sutherland, *et al.* (1986) conducted the same test, this time using life expectancy *from age at maturity* instead of life expectancy at birth. The correlation controlling for body size remained significant and the two body-size-corrected measures are plotted against each other in Fig. 5.6.

Another example of a higher nodes approach comes from an analysis of the hippocamapal complex in birds reported by Sherry *et al.* (1989). Some bird species collect and store large numbers of food items, each in a separate place, before retrieving them at some later date. In contrast, many other bird species immediately consume the food that they gather.

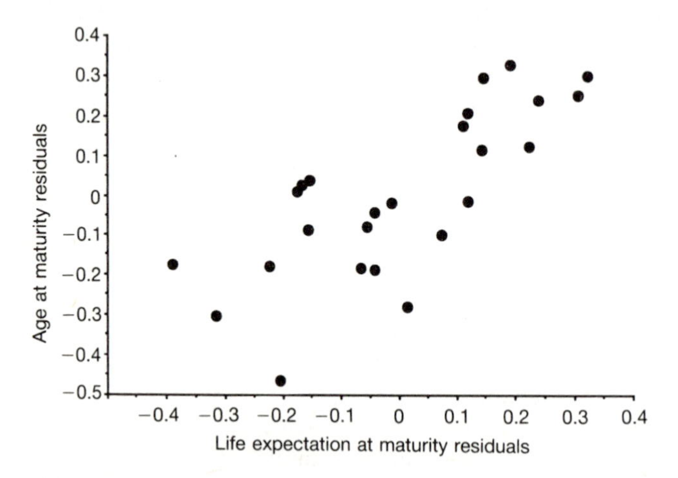

Fig. 5.6. The deviations from the logarithmic regression of age at maturity on body weight plotted against the deviations from the regression of life expectation at maturity on body weight. A positive deviation indicates that the life history variable was larger than would be expected on the basis of body weight, a negative deviation indicates the opposite. The two size corrected measures are highly correlated (r= 0.76, P<0.001). The data points are generic averages of constituent species values.

Sherry and his colleagues were interested in the idea that storing and retrieving food places demands on the memory of food-storing species not experienced by species that do not store their food. They suggested that food-storing birds might have enlarged hippocampuses as a result, because the hippocampal complex is thought to be involved in spatial memory.

Food-storing versus non-food-storing does not vary independently among species belonging to different families of birds. For example, many North American chickadees (Paridae), nuthatches (Sittidae), and jays and crows (Corvidae) store food while species in other passerine families and subfamilies do not. Sherry *et al.* (1989) measured the size of the hippocampal complex for 23 species from 13 passerine subfamilies, and performed their analysis across subfamily averages. They justified this by the fact that ten of the subfamilies in the data set were represented by only non-food-storing species while the other three subfamilies were represented by only food-storing species. The food-storing subfamilies had large hippocampal complex sizes relative to both their body weights and to the volumes of their telencephalons, which is the part of the brain within which the hippocampus complex is situated. Figure 5.7(a) plots body-size-corrected hippocampus complex volumes for each subfamily of birds in Sherry *et al.*'s data set.

Krebs *et al.* (1989) were able to use an interestingly expanded data set to tackle the same problem. Two of the food-storing families, the Paridae and the Corvidae, contain some species that do not store food. Krebs *et al.* measured the hippocampus size of species that store food and those that do not store food within each family. Do the non-food-storing members of these families have relatively smaller hippocampuses than their food storing relatives? They do, as can be seen in Fig. 5.7(b). Furthermore, the Troglodytidae which are non-food-storers have relatively smaller hippocampuses than the closely related food-storing Sittidae. Krebs *et al.*'s analysis goes beyond comparing higher taxon means by also examining variation within higher taxa.

What assumptions about evolution and the phylogeny of species are embedded in the higher nodes method? The method treats the higher nodes as independent points in analyses. Technically, this means that the higher nodes must have a phylogeny that forms a simultaneous radiation pattern like that in Fig. 5.1. Thus, for example, if families were the unit of analysis, the phylogeny must be such that all of the families simultaneously radiate from a single ancestor common to the entire class. All branches leading to families must be the same length to ensure equality of expected variances. Branch lengths can be ignored if the amount of evolution is believed to have been independent of time and yet equal in each branch (e.g. punctuational change). The branches leading to orders must have a length of zero: any order branch with a non-zero length would possibly

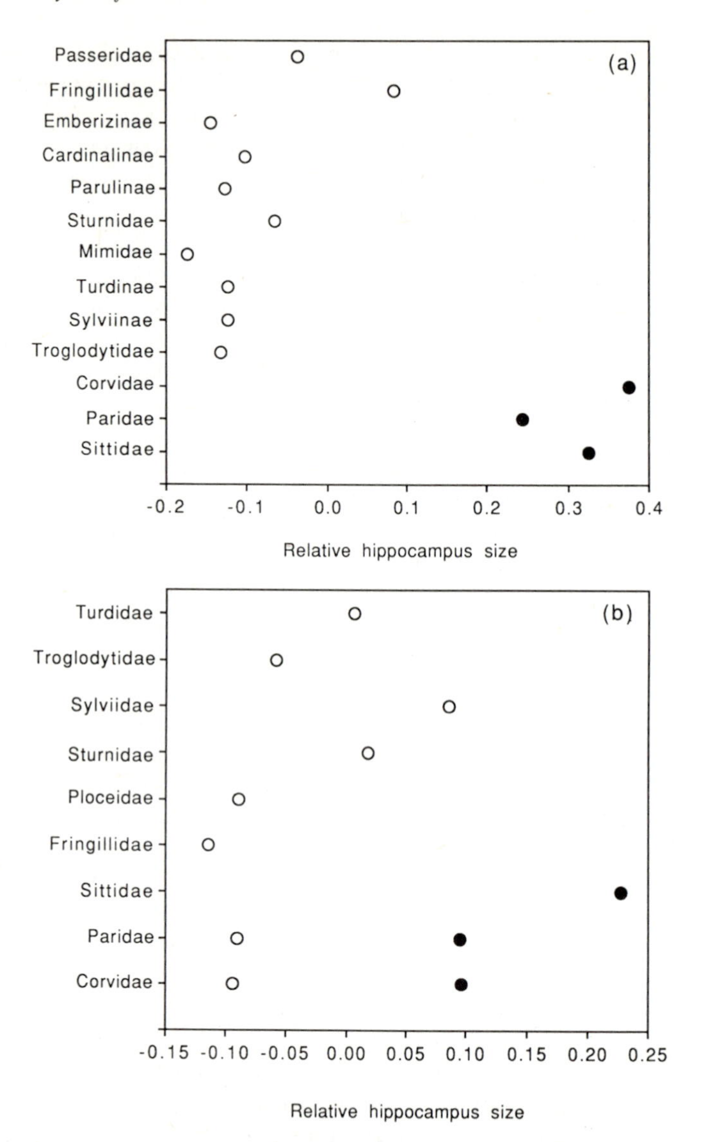

Fig. 5.7. The higher nodes approach illustrated by hippocampal complex size in passerine birds. (a) When body size effects are controlled for (relative hippocampus size is the deviation from the cross-subfamily regression of hippocampus volume on body weight), food-storing subfamilies of birds (●) have relatively larger hippocampal complexes than their non-food-storing relatives (○). (Data from Sherry *et al*. 1989). (b) When body size effects are controlled for, non-food-storing parids and corvids have relatively smaller hippocampuses than their food-storing relatives. Furthermore, the food-storing Sittidae have relatively larger hippocampuses than the closely related non-food-storing Troglodytidae. (Data from Krebs *et al*. 1989).

introduce phylogenetic similarity among its families (there are other models in which the families could be independent but this is the simplest and most general). Further assumptions are built into the family means themselves. By finding genera means first, and then family means as the average of their respective genera, all lower taxa are assumed to be simultaneous radiations with equal branch lengths.

It would be very unusual for any phylogeny to have the form required for the higher nodes method to be correct technically. In practice, however, this does not mean that conclusions drawn from all higher nodes analyses are incorrect. The most obvious worry in using a higher nodes approach is that the higher nodes may not be independent for the same sorts of reasons that species are not independent. One should, for example, be careful to examine the data to see if the result depends upon the contribution of any one cluster of points that share an immediate ancestor. The relationship can also be examined separately within taxonomic groups. For example, Krebs *et al.* (1989) were able to show that in all three cases where pairs of closely related taxonomic groups could be compared, the hippocampus volume was greater in the food-storing group. However, even if all of the assumptions of the higher-nodes method were met, it would still have the unavoidable limitation that information from lower taxonomic levels is lost, along with degrees of freedom for statistical tests.

5.5 Stearns' phylogenetic-subtraction method

A method for directly subtracting from the species' data the variation associated with phylogenetic similarity (or taxonomic similarity where taxonomy is used to stand in for phylogeny) was developed by Stearns (1983). Stearns' method, although motivated by the same concerns as the higher nodes method, proceeds in a manner opposite to it.

Stearns proceeds on the assumption that the portion of the total variation associated with differences among higher nodes represents lineage-specific variation that is not appropriate for testing questions about adaptation at the species level. Stearns statistically removes from the data the lineage-specific variation associated with higher nodes, and analyses the remaining variation. Stearns' (1983) analysis of mammalian life histories provides an example. Differences among orders and among families within orders were assumed to represent effects of phylogenetic similarity present in the species data points. Stearns simply subtracted from each species data point the mean value for its order. Species' values were then free of similarity associated with differences among orders. This procedure was repeated using family means, thus leaving a set of residual data points for species that were free of differences associated with families or orders. Stearns found that substantial covariation existed among mammalian life history variables even after removing these taxonomic associations. Wooton (1987) employed a similar approach in his study of age at first reproduction in mammals.

Stearns' nested analysis of variance method for controlling phylogenetic effects is statistically equivalent to removing the variance using categorical codes in a multiple regression. Box 5.1 gives an example of the categorical coding required to remove order, family, and generic differences among sixteen species in a hypothetical symmetrically bifurcating phylogeny such as that in Fig. 5.3. If the *Y* variable (and any other variable to be analysed) is regressed onto the categorical codes, and residuals are found, these residuals will be statistically independent of their phylogeny. This has been done for a hypothetical data set in Box 5.1.

Box 5.1. Removing phylogenetic correlations from a continuously-varying character using multiple regression

S	O	F	G	O1	F1	F2	G1	G2	G3	G4	*Y*	Residual
1	1	1	1	1	1	0	1	0	0	0	29	2.0
2	1	1	1	1	1	0	1	0	0	0	25	-2.0
3	1	1	2	1	1	0	0	0	0	0	26	-1.5
4	1	1	2	1	1	0	0	0	0	0	29	1.5
5	1	2	3	1	0	0	0	1	0	0	24	3.5
6	1	2	3	1	0	0	0	1	0	0	17	-3.5
7	1	2	4	1	0	0	0	0	0	0	24	-2.5
8	1	2	4	1	0	0	0	0	0	0	28	2.5
9	2	3	5	0	0	1	0	0	1	0	15	1.0
10	2	3	5	0	0	1	0	0	1	0	13	-1.0
11	2	3	6	0	0	1	0	0	0	0	14	-2.5
12	2	3	6	0	0	1	0	0	0	0	19	2.5
13	2	4	7	0	0	0	0	0	0	1	12	4.5
14	2	4	7	0	0	0	0	0	0	1	3	-4.5
15	2	4	8	0	0	0	0	0	0	0	14	3.0
16	2	4	8	0	0	0	0	0	0	0	8	-3.0

Sixteen species are classified by phylogenetic relatedness into order (O), family (F) and, genus (G). The taxa are then given categorical codings to produce dummy variables, one for order membership (O1), two for family membership (F1, F2), and four for genus membership (G1, G2, G3, G4). Each categorical code is used as an independent variable in a hierarchical multiple regression (Draper and Smith 1981) to remove the taxonomic correlates of a continuously varying character (*Y*), resulting in a set of residuals that are not correlated with taxonomy. Thus, code O removes the differences between the means of the two orders, code F1 removes differences among the families 1 and 2 nested in order 1, and so on down to code G4 which removes differences between genera 7 and 8 nested in family 4.

The correlation between the residuals and any one or combination of the discrete phylogenetic dummy variables in Box 5.1 will always be exactly zero. Stearns (1983) removed taxonomic similarity due to orders and families. However, as Box 5.1 shows, it is possible to remove the effects of taxonomy down to the genus level, or more generally, down to one level above the lowest level represented in the data set. This procedure for removing variation associated with phylogeny can be applied to any number of variables, and then the relationships among them tested.

Returning to the conceptual model outlined at the beginning of this chapter, what phylogenetic structure is implicit in Stearns' approach? Stearns' approach assumes that, below some level (Stearns chose families), the phylogenetic or taxonomic groups are independent. The simultaneous radiation of Fig. 5.1 produces independence, but imagination is required to see how this model can be applied to groups that cannot be thought of as sharing an immediate common ancestor. The effect of removing variation associated with phylogeny is to make the mean value of the trait equal to zero within the lowest level groups controlled for in the analysis. Thus, in the example in Box 5.1, the mean value of the residuals within genera is exactly zero. In this statistical sense, then, the species can be thought of as having a common phenotypic starting point of zero. In Stearns' life history study, the family level was the lowest level controlled for, and so the implicit phylogeny must have all species in all genera radiating from a common starting point.

We have applied Stearns' approach to the Millar and Zammuto (1983) data set to illustrate the difference between this approach and the higher-nodes method. The data set contains six different orders and 18 families. Harvey and Zammuto (1985) and Sutherland *et al.* (1986) conducted their analyses of life expectancy, age at maturity, and body size across genera means ($n = 25$). We used multiple regression to remove the variation in these three variables that is associated with order and family differences among the 29 species. Then we repeated Harvey and Zammuto's analyses on the species data with order and family effects removed.

The dummy coding was done in a manner analogous to that in Box 5.1. Five dummy codes were required to control for the six orders. Only six additional codes were required to control for differences among families within orders because of the way that families were distributed among the orders. With additional codes for genera within families and species within genera, it is possible to reconstruct the nested analysis of variance table for this data set. The species are taxonomically diverse in this data set, so many families are represented by just one or two species. Because of this, the phylogenetic differences among orders and families account for a very large percentage of the total variation among the species.

Table 5.2 Taxonomic distribution of variance among the characters used for the analysis of life history variation and mortality among placental mammals. Tabulated values are percentage of total variance accounted for at successive taxonomic levels estimated from a nested ANOVA on logarithmically transformed species averages. (Data from Millar and Zammuto 1983).

Among: Within:	species genera	genera families	families orders	orders class
Variance component:*	$\sigma^2_{s(g)}$	$\sigma^2_{g(f)}$	$\sigma^2_{f(o)}$	$\sigma^2_{o(c)}$
Body weight	1	2	12	85
Age at maturity	1	3	36	60
Life expectation at maturity	2	2	34	62

*Following eqn (5.2), each variance component in the table has been multiplied by 100 and divided by σ^2_{tot}

By using just the dummy codes for orders and families, the 29 species data points are statistically independent of phylogenetic variation associated with those levels. We then further controlled both life history variables for body size, and plotted the residuals against each other. Fig. 5.8 plots this relationship which although positive is not as strong as that in Fig. 5.6. Controlling for phylogenetic relationships and for body size, there is no longer a significant relationship between age at maturity and life expectation at maturity ($r = 0.35$, d.f. = 16, $P > 0.15$; note that the degrees of freedom are equal to the sample size minus the number of control variables minus 1: 29–12–1).

An investigator using Stearns' method on the Millar and Zammuto data set would have reached a different conclusion from that which Harvey and Zammuto (1985) and Sutherland *et al.* (1986) reached using a higher nodes approach. Which would be correct? Both would be correct for certain kinds of conclusions (Pagel and Harvey 1988*a*). The conclusion that there is not significant covariation between the life history variables after controlling for differences associated with phylogeny is correct for this data set. So is the conclusion that there is significant variation across genera. The interesting debate concerns what we should make of the difference.

Variation among higher nodes is removed by Stearns' method and variation at lower levels is retained. The higher-nodes method analyses the variation at higher levels and averages over the lower levels. Each method uses the information that the other method discards! Stearns assumed that

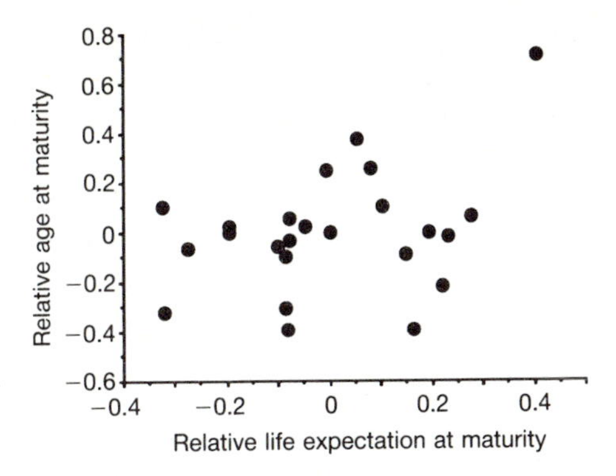

Fig. 5.8. The relationship between age at maturity and life expectation at maturity for mammals after controlling for body weight and phylogenetically correlated variation in each variable. The correlation is not significant [r=0.35, d.f.=16 (see text), P>0.15]. 29 species are represented on the figure, but 6 data points overlap at 0,0.

variation at higher levels should be attributed to lineage-specific trends and thus was inappropriate for testing questions about function. His point is that the adaptive variation is that which is independent of differences associated with phylogeny. The investigator using Stearns' method should explain why variation among higher taxonomic groups is thought to be inappropriate for testing this functional question. The higher-nodes investigator must provide an explanation of why it is felt that the differences among the higher taxonomic levels are not confounded by other taxonomic differences.

Consider, for example, that all of the rodents have early ages at maturity and short life expectancies from maturity not because of a causal connection between the two, but for some other reason associated with being a rodent. This would represent a taxonomic confound that the Stearns' method would remove. Thus, the higher nodes investigator using the order as the level of analysis should demonstrate that the relationship is found within orders, like the rodents, as well as across them.

Both sides of this debate have something to offer. However, we shall show in a later section how it is possible to get around the conflict between these two methods by using techniques that analyse the covariation between traits within taxonomic (or phylogenetic) groups. There is no need to throw out the variance at either the lower or the higher levels. All of the variation in the data can be used to test ideas about the correlation between traits.

5.6 Phylogenetic autocorrelation method

Cheverud *et al.* (1985) describe a method for partitioning phenotypic traits into phylogenetic and specific components that is conceptually similar to Stearns' method. Cheverud *et al.* predict a species' phenotype on the basis of the phenotypes of other species in the sample. More closely related species, such as those in the same genus, will typically be better predictors than more distantly related ones. Accordingly, the phenotype of a focal species that is closely related to many other species in the sample will be better predicted than one that does not have many close relatives. This fits with our intuitive feeling that closely related species do not each represent an independent instance of the evolution of their trait. However, species' phenotypes will not be perfectly predicted, even when they share many close relatives. Cheverud *et al.* (1985) use this *specific* portion of the phenotype to test for correlated relationships among variables. Their method, then, although conceptually similar to Stearns' (1983), differs by employing an explicit evolutionary model to estimate variation due to phylogenetic effects.

Cheverud *et al.*'s phylogenetic autocorrelation method uses a linear autocorrelation model to partition the total variance in a trait that is measured across species into the sum of phylogenetic and specific variances, plus the covariance between the phylogenetic and specific values of the trait. The model represents the trait y as a linear combination of phylogenetic and specific effects according to:

$$y = \rho W y + e. \qquad (5.3)$$

where y is the vector of length n containing the n species' data points, ρ is a scalar 'phylogenetic autocorrelation coefficient', W is an $\mathrm{n} \times \mathrm{n}$ 'phylogenetic connectivity matrix', ρWy is a vector of predicted y values representing the phylogenetic portion of y, and e represents the vector of residual values of the trait that cannot be predicted by the vector ρWy. It is e that is used to analyse whether there is covariation between traits that is independent of phylogenetic effects. Cheverud *et al.*'s (1985) model assigns both parallel evolution and variance due to the interaction of the phylogenetic and specific components solely to the phylogenetic effect.

The matrix W is used to account for the phylogenetic relatedness of species. The phylogenetic portion of the attribute y, given by the vector ρWy, is just a weighted sum of the phenotypic trait values of each of the other species in the data set, scaled by the factor ρ. The scalar quantity ρ is roughly a measure of the correlation between the observed and predicted values of y, where the predicted values are equal to ρWy. Thus, if the actual values of y are largely predictable from phylogenetic relatedness, then ρ will be high.

The weights in W are assumed to be big for closely related species, and to decline for more distantly related species. In a worked example, Cheverud *et al.* (1985) arbitrarily assigned weights of 1.0 for congeners, 1/2 for species in the same family, 1/3 for the same superfamily, and 1/4, 1/5, and 1/6 for the same infra-order, suborder, and order, respectively. Thus, a trait value for a species that is closely related to a large number of other species in a data set is attributed primarily to phylogeny. The authors discuss several ways of estimating the relatedness weights; Gittleman and Kot (1991) report a method that allows an assessment of the weighting according to the variance in the data.

Cheverud *et al.*'s (1985), method can be illustrated for the phylogeny of Fig. 5.9.

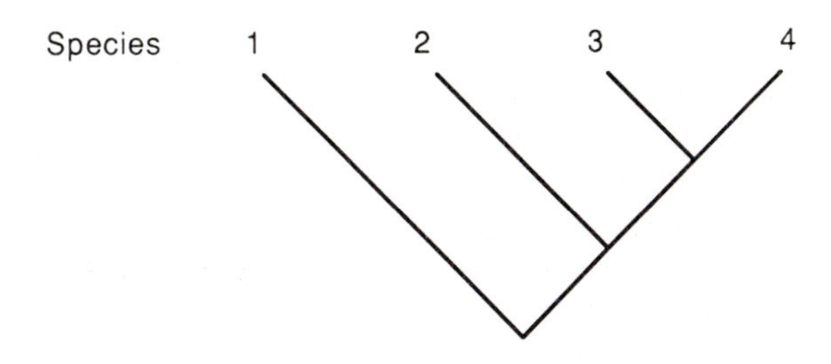

Fig. 5.9. A phylogeny used to illustrate Cheverud *et al.*'s (1985) phylogenetic autocorrelation method. Species 1, 2, 3, and 4 possess the character values 10, 8, 3, and 4 respectively.

We assume that the four species comprise a single order. Species 3 and 4 are congeners, species 1 is in a different family from 2, 3 and 4. A hypothetical reconstruction of the terms in eqn (5.3) is given below. This example is for illustrative purposes only and should not be taken to indicate the actual values that Cheverud *et al.*'s procedure would produce.

$$\begin{array}{ccccc} y & = \rho & W & y & + \quad e \\ \begin{bmatrix} 10 \\ 8 \\ 3 \\ 4 \end{bmatrix} & = [0.49] & \begin{bmatrix} 1 & 1/6 & 1/6 & 1/6 \\ 1/6 & 1 & 1/2 & 1/2 \\ 1/6 & 1/2 & 1 & 1 \\ 1/6 & 1/2 & 1 & 1 \end{bmatrix} & \begin{bmatrix} 10 \\ 8 \\ 3 \\ 4 \end{bmatrix} & + \begin{bmatrix} 3.88 \\ 1.55 \\ -3.21 \\ -2.21 \end{bmatrix} \end{array}$$

and thus

$$
\begin{matrix} y & = & \rho Wy & + & e \\ \begin{bmatrix} 10 \\ 8 \\ 3 \\ 4 \end{bmatrix} & = & \begin{bmatrix} 6.12 \\ 6.45 \\ 6.21 \\ 6.21 \end{bmatrix} & + & \begin{bmatrix} 3.88 \\ 1.55 \\ -3.21 \\ -2.21 \end{bmatrix} \end{matrix}
$$

This example shows how the matrix product ρWy produces a predicted y score for each species, and the vector e is just the residual difference between y and the predicted y. Species that are closely related to each other have a greater amount of their trait attributed to phylogeny. Species 3 and 4 have the same predicted value of y because they are congeners. Our illustration uses an arbitrary value for ρ that makes the predicted values have the same mean as the actual values. In practice, Cheverud *et al.* use a maximum likelihood procedure for estimating ρ.

Cheverud *et al.* (1985) used their model to study body size dimorphism in primates. Four variables, body size, mating system, habitat, and diet, were studied for their association with the extent of sexual dimorphism in body size for 44 primate species. The phylogenetic autocorrelation procedure was used to define phylogenetic (ρWy) and specific (e) components for each trait. Then the eight components plus the phylogenetic portion of size dimorphism were used to predict size dimorphism. Table 5.3 shows the unique contribution of each variable in terms of the percentage of the variance in size dimorphism that it predicted.

Table 5.3 Proportions of total variance in sexual body size dimorphism explained by phylogenetic and specific sources of variance (after Cheverud *et al.* 1985). *Note*: inclusion of the specific proportion of sexual dimorphism would (arbitrarily) have led to 100 per cent of the variance being accounted for.

	Percentage of variance accounted for by sources of variance	
Trait	Phylogenetic	Specific
Mating system	<1	<1
Body weight	28	34
Sexual size dimorphism	1	–
Habitat	2	4
Diet	<1	2
Interaction among traits	19	2
Total	50	42

Fifty per cent of the variance in size dimorphism can be accounted for by the phylogenetic components, and an additional 42 per cent by the specific components.

As with Stearns' procedure, potentially the majority of the variation is assigned to phylogenetic components and is treated as non-adaptive. Investigators should be aware of this and provide an explanation of why they feel that variation among higher taxonomic groups should not be used to test questions about function and adaptation.

5.7 A maximum likelihood approach

Lynch (in prep.) reports a method that, like Cheverud *et al.*'s (1985) method relies on a statistical model to partition each species' phenotypic trait value into components associated with and components independent of phylogeny. Unlike Cheverud *et al.*, however, it is the phylogenetic component that Lynch uses to test the comparative relation.

Lynch borrows ideas from quantitative genetics to partition species' phenotypic values into three components, two representing phylogenetic variation, and one representing variation that is independent of phylogeny. Each species' phenotypic mean is seen as a combination of an overall phylogenetic effect, a component representing the 'heritable additive evolutionary value of the character', and a residual component. The overall phylogenetic effect is analogous to a grand mean on the trait. The interpretation of the additive heritable component is that it represents something akin to a breeding value: a species' additive effect is the expected phenotype of a descendant of that species. The sum of the first two components is the overall 'heritable component of a particular realization of the evolutionary process'. The residual effect represents non-additivity of genetic effects, environmental effects, and sampling error.

Knowledge of the across-species variance-covariance matrices of the additive effects and of the residual errors is required to estimate the additive effects and the overall effects. The matrix of additive effects, in turn, depends on a matrix that measures the true phylogenetic relationships among taxa. For example, all species are perfectly related to themselves, less so to sister taxa, and so on, in a fashion similar to Cheverud *et al.*'s (1985) matrix W. The phylogenetic relationships matrix is estimated from the phylogeny. The relatedness of two species is taken as the proportion of their total path lengths that they share. The additive effects and the overall effects are estimated by a recursive maximum-likelihood procedure. Using initial arbitrary values of the additive effects and residual errors, the variance-covariance matrices can be found. This, then, leads to new estimates of the additive and residual effects, and so on until a stable solution is reached. Lynch reports that the maximum-likelihood algorithm usually converges, or at least leads to a region of results. Statistical convergence may also be stymied by multiple peaks in the likelihood surface.

Lynch's method, by using explicit statistical criteria to take into account the non-independence of taxa, has much to recommend it. Like Cheverud *et al.*'s (1985) technique it avoids altogether the problem of reconstructing ancestral character states, instead conditioning all tests on the variation among extant species. It is difficult to judge at this point how well Lynch's approach will work, and how it will manage with poorly known phylogenies. It is critical to get the estimates of the variance-covariance matrices correct in order to adjust the species additive effects for their strong phylogenetically based lack of independence. It may also be unnecessary, as we shall suggest below, to partition the species' phenotypes as Stearns (1983), Cheverud *et al.* (1985), and Lynch do. Nevertheless, maximum-likelihood methods such as Lynch's deserve more attention.

5.8 Independent comparisons methods

All of the methods reviewed so far, with the exception of a species regression, make a distinction between variation associated with phylogeny, and variation that is independent of phylogeny. The methods to be described in this and the following sections use all of the variation in a trait to test for a comparative relation, and they do so without partitioning the traits into phylogenetic and non-phylogenetic components. Independent-comparisons methods are able to make use of all of the data by recognizing that what is phylogenetic inheritance at one level of a hierarchy may constitute part of an adaptive difference at the next highest level.

This discussion is based on logic outlined by Felsenstein (1985*a*). Figure 5.3 shows a branching phylogeny for eight species. Focussing on a portion of this phylogeny, the range of values including the two species that split from node *n*4 and the two species that split from node *n*5 is largely the result of a phylogenetic difference that evolved once between *n*4 and *n*5. That is, most of the variation among these four species in a typical phylogeny would have already been present between the two higher nodes. However, there are three degrees of freedom among these four species: the difference between species 1 and 2, the difference between species 3 and 4, and the difference between nodes *n*4 and *n*5.

Assume that changes along the branches of the phylogeny can be modelled by a Brownian motion process such that, as above, successive changes are independent of one another, and that the expected total change summed over many independent changes is zero. Then, the three pairwise differences (species 1 versus 2, species 3 versus 4, and *n*4 versus *n*5) are independent of each other. This is because, for example, the difference between species 1 and 2 reflects only the evolutionary changes that have taken place since they split from their common ancestor (*n*4). All similarity between species 1 and 2 that is due to their shared phylogenetic history will be, in effect, subtracted out. The same logic applies to species 3 and 4. Their difference in turn will be independent of the differences between species 1 and 2. Finally, the difference between nodes *n*4 and *n*5 reflects only the evolutionary events that have happened since they split from their common ancestor, and this difference will be independent of the other two.

The three comparisons together account for all of the variation among the four species by dividing the variation into three separate evolutionary events. Each event reflects the difference between the evolutionary changes in two branches of the tree. So, the advantage of independent-comparisons approaches is that, by partitioning the variation appropriately, it can all be used to assess the comparative relationship.

More generally, in a given branching phylogeny we might calculate the difference in Y and difference in X between the species within each of the lowest level clades, then again at the next highest level, and so on until we compare the two highest nodes of the tree. The important point is that each of these relationships represents under the null hypothesis an independent instance of the evolution of the relationship between Y and X. Thus, any covariation between Y and X that is present among the species sharing a common ancestor is phylogenetically independent of the covariation between Y and X among the species sharing a different common ancestor. The same argument applies to each similarly defined pair of higher nodes. The set of differences between Y and X provides a way to test whether changes in Y and X are correlated. Under the null hypothesis that evolutionary changes in Y and X are unrelated, a positive difference on X should be associated with a positive difference on Y no more often than with a negative difference. A preponderance of positive (or negative) relationships within taxa, then, is evidence against the null hypothesis.

Box 5.2 provides a simplified summary of the procedure used to produce and compare independent comparisons. Note, however that path lengths are ignored in the example and that higher nodes are calculated as the average value of lower nodes. We shall discuss both of these issues later.

Later in this section we shall discuss three methods that use independent comparisons. The methods all rely on the same independent-comparisons logic but differ in assumptions and statistical manipulation of the data. Before discussing the three independent-comparisons procedures we make a brief digression to describe the nested analysis of covariance.

5.8.1 A brief digression: nested analysis of covariance

The nested analysis of covariance does not make use of the logic of independent comparisons, but it does exploit the fact that species naturally form a nested hierarchy. Following brief sorties by Dunham and Miles

Box 5.2. The independent comparisons method for two characters in a single phylogeny.

Under a Brownian motion model of evolution, d1, d2, and d3 provide independent comparisons. Path length differences are ignored in this illustration.

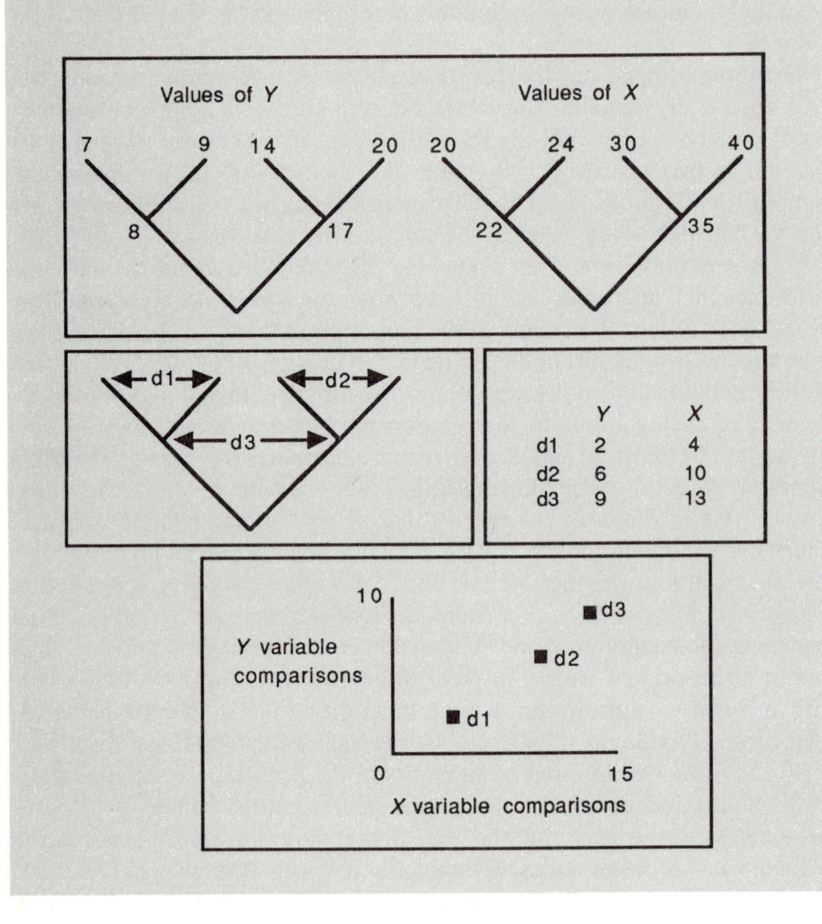

(1985) and by Martin and Harvey (1985), Bell (1989) suggests a novel use of nested analysis of covariance to analyse comparative data.

A nested analysis of covariance analyses the covariation, or equivalently, the correlation between two or more variables separately within each of the groups in a nested hierarchy. Applied to comparative data, the method finds a relationship between two or more variables separately within each taxonomically or phylogenetically defined group in a data set.

The separate within-group relationships can then be combined to form a pooled estimator of the within-group correlation between two variables. A pooled estimator of the correlation between two variables is analogous to a pooled estimator of the within-groups variance in an ordinary analysis of variance, except information about the covariation between two variables rather than information about the variation of one variable is used.

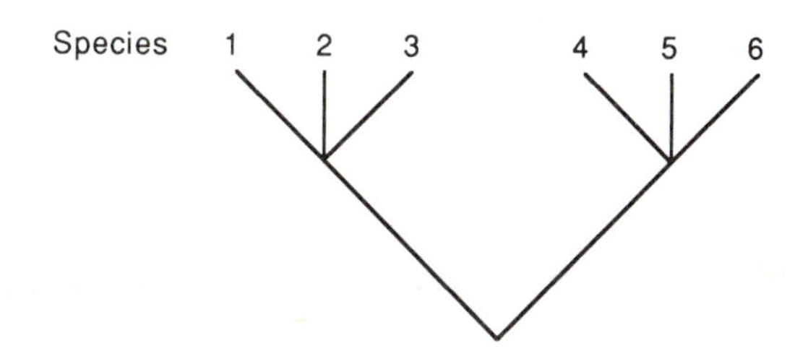

Fig. 5.10. A phylogenetic tree used to illustrate the nested analysis of covariance method.

Consider the hypothetical tree of Fig. 5.10 showing three species in each of two genera. Ignoring the genera classifications for a moment, let $\sigma_{\mathrm{P}xy}$ represent the phenotypic covariance between Y and X across the six species. The covariance is defined as the correlation between X and Y multiplied by the standard deviations of X and Y:

$$\sigma_{\mathrm{P}xy} = \rho_{\mathrm{P}xy}\sigma_{\mathrm{P}x}\sigma_{\mathrm{P}y}. \tag{5.4}$$

Now, classifying the species into their genera, it is possible to partition the phenotypic covariance into two components, one representing the covariation between Y and X within genera, and one representing their covariation across genera:

$$\sigma_{\mathrm{P}xy} = \sigma_{\mathrm{W}xy} + \sigma_{\mathrm{A}xy} \tag{5.5}$$

where the subscripts 'W' and 'A' refer respectively to 'within' and 'among'. The covariance within genera is the sum of the separate covariances in each of the two genera. The individual within-genus covariances are not influenced by phylogenetic differences among the species if the genus is a monophyletic group. The covariance among genera is calculated across the two genera means. This covariance expresses a phylogenetic difference between the two groups of species. Thus the overall covariance among the species is a combination of phylogenetic and non-phylogenetic or evolutionary components. The within-groups covariance represents an evolutionary relationship between Y and X that is not influenced by differences in phylogeny among the species. This is true because, if Fig. 5.10 is an accurate representation of the true phylogeny, species 1 through 3 share their phylogenetic history as do species 4 through 6. This method can be applied to all levels of a phylogeny. For example, a pooled estimator of the covariance between Y and X calculated from genera means within families, has an analogous interpretation to the covariance among species within genera. Similar partitionings can be made at higher levels.

Bell (1989) applied the nested analysis of covariance method to study the relationship between litter mass and gestation length in mammals. Both variables are highly correlated with body weight in mammals and so, as a first step, Bell statistically removed the association of both variables with body weight. All analyses were then conducted on relative litter mass and relative gestation length, where the relative values are defined as the residuals from the respective regression lines for the two variables on adult body weight. Bell compiled 574 observations of litter mass and gestation length on 370 species, then conducted a nested analysis of covariance on the two body-weight-corrected measures. The results are given in Table 5.4 where, for ease of interpretation, Bell has converted each of the covariances to a correlation by dividing by the product of the two standard deviations. The correlations in Table 5.4 are found by pooling all of the individual within-group correlations at a given taxonomic level.

Table 5.4 Nested analysis of covariance results for relative litter mass and relative gestation length in 370 mammal species. The correlations are estimated for the taxonomic level indicated and are nested within the next highest level. For example, the within-species correlation is that for individuals within species. NE = not estimated.

Taxonomic level	Degrees of freedom	Estimated within-group correlation
Superorder	4	NE
Order	12	NE
Suborder	6	0.69
Family	52	0.52
Subfamily	47	−0.01
Genus	120	0.21
Species	128	−0.10
Within species	204	0.33

Bell was able to calculate a pooled correlation among individuals within species because more than one observation was available for many of the species. The results show substantial correlations within species, and then again among families within suborders, and suborders within orders.

The main difference between Bell's approach and the methods reviewed previously in this chapter is that Bell's analysis makes use of all of the variation in the data set to assess the comparative relation. Variation among species within genera, genera within families, and so on to the highest nested level is all used to investigate whether two variables are correlated. This method is very close to what we have called 'independent comparisons methods' in that it avoids phylogenetic influences by looking separately within taxonomically or phylogenetically defined groups. Provided that all of the members of a group share an immediate common ancestor, there is no phylogenetic variation within the group, and the correlation between two variables within that group represents evolutionary change since they diverged from their common ancestor. If the groups are not monophyletic then some phylogenetic differences will be included in the within-groups covariance.

Combining the information from different taxonomic groups and levels requires assumptions about phylogenetic branch lengths (assuming a fixed gradual model of change). The branch lengths leading from each common ancestor to their respective descendant taxa must be the same length in each group. If they are not, then the expected variation within a taxonomic group with longer branch lengths will be larger than that within a taxon having short branch lengths. Bell combines information only from the same taxonomic level, thus, having only to assume that all taxa at the same level have equivalent branch lengths.

The pooled estimators combine information from different groups. This gives more weight to groups that have more subtaxa. For example, a genus with seven species will contribute a larger share to the pooled estimator than a genus with two species. Such a weighting would be appropriate if the true phylogeny within the genus was a real simultaneous radiation of the n subtaxa. Then, there would be n–1 independent pieces of information among them. But consider that the true but unknown phylogeny is not a simultaneous radiation. Then the n subtaxa will not be independent and treating them as such will spuriously give more weight to larger groups (in the limiting case of no phylogenetic classification whatsoever, the nested analysis of covariance method would be identical to a species regression, with all species nested within a single higher node). Ideally, we should like to know how to weight such groups where we have reason to believe that the n subtaxa are not independent, but at the same time the true phylogeny is unknown.

The methods that we discuss in the following sections address the problem of weighting by allowing each taxonomic group to contribute only one piece of information that is not weighted by the number of subtaxa. We shall now discuss three methods that do this, each based on a different evolutionary model.

5.8.2 Felsenstein's method of pairwise independent comparisons

Felsenstein (1985*a*) was the first to develop a method for testing comparative relationships based on the logic of comparing pairs of species or higher nodes that share a common ancestor (Section 5.8). Felsenstein's method is based on a Brownian motion model (Section 5.2) of evolutionary change. Other models of change are possible, although most are intractable statistically (Felsenstein 1988). The method requires that the true branching phylogeny, including the lengths of the branches of the tree in units of expected variance of evolutionary change, is known. This information allows the calculation of a set of comparisons among pairs of data points, each of which has the same expected mean and variance under the null hypothesis.

Consider the phylogeny in Fig. 5.11. Following the logic developed above for independent comparisons methods, and drawing on Felsenstein's (1985*a*) article, the differences between the two pairs of species at the tips of the tree, represented by $(X1-X2)$, and $(X3-X4)$, will be independent of each other. By the same logic the difference between the higher nodes defined by $(X1+X2)/2$ and $(X3+X4)/2$ is another comparison, and it is independent of the first two. (Here we assume that $v1=v2=v3=v4$, and that $v5=v6$. If path lengths vary the comparisons change somewhat. Felsenstein's original paper gives formulae for the general case). Calculating the comparisons among the higher nodes this way ensures, given the Brownian motion model, that the comparisons are mutually independent

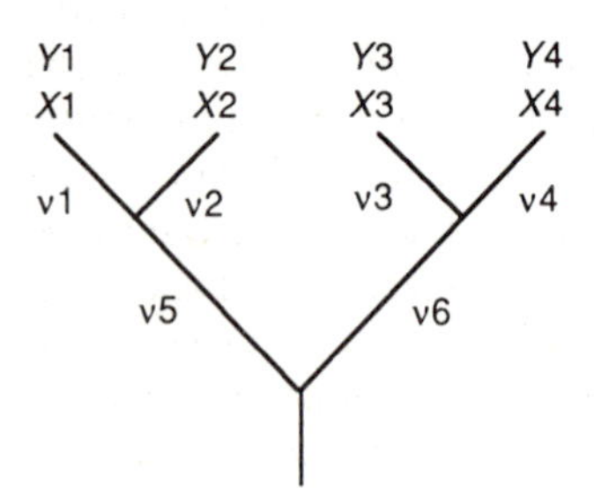

5.11. A phylogeny of four species. X_i and Y_i (i = 1–4) represent the states of phenotypic characters in species i. The v_j (j = 1–6) define the time in units of expected variance of evolutionary change spent evolving along each branch of the tree. (After Felsenstein 1985*a*).

statistically. Calculating the same three comparisons for the Y_i would yield three pairs of points that could be used to ask whether changes in the X variables go with changes in the Y variables. In general, with n species, we can find a set of $n-1$ mutually independent comparisons in a bifurcating tree.

Before discussing the use of the independent comparisons we should point out why such comparisons are used rather than those that measure the independent evolutionary change along each branch (which we referred to in Chapter 1 as 'directional' methods). We could use the latter if we had a method for reconstructing ancestral conditions that was independent of the descendant character states. But consider that, in Felsenstein's method for example, the higher nodes are roughly equal to the arithmetic means of the lower nodes. As the sum of the deviations of a set of scores around their mean must sum to zero, this means that we cannot use both scores for a pair of species. But we can use their difference.

The set of differences are independent under the null model, but they will not all have the same expected variance. Here is where the branch lengths and Felsenstein's evolutionary model are put to use. Felsenstein models the evolution of a character along its branch by a process of Brownian motion. If change is independent in each unit of time, then after one unit of time the character will have accumulated σ_x^2 units of variance, where σ_x^2 is the variance of the process (For ease of discussion, we shall assume that σ_x^2 is constant throughout the tree. But it need not be.). Then, over v units of time the variance will be $v\sigma_x^2$. This means that the various observations on X will have the same variance only if their branches are of the same length.

However, having knowledge of the variance makes it possible to scale each X score to have a mean of 0 and standard deviation of 1:

$$\frac{(X-0)}{\sqrt{\upsilon\sigma_X^2}} \qquad (5.6)$$

The same calculations can be performed for the Y variables also using the $\upsilon\sigma_y^2$ (however σ^2 for the X variable need not be equal to σ^2 for the Y variable). Then, each difference between a pair of species or higher nodes will also be a variate with a mean of 0 and a standard deviation of 1. If the evolution of the characters can be described by Brownian motion, then the set of comparisons on X and Y can be regarded as having been drawn from a bivariate normal distribution with means of 0, standard deviations of 1, and unknown correlation parameter, ρ. The null hypothesis is that ρ equals 0.

Sessions and Larson (1987) used Felsenstein's method to test whether genome size in plethodontid salamanders is related to developmental rate. The 'junk DNA' hypothesis predicts that junk DNA will accumulate in the genome until the costs of transcribing it impose too great a cost on the organism. This leads to the prediction that genome size (as measured by the *C*-value, the weight of the genome in picograms) should be inversely related to measures of the developmental rate of a species.

Sessions and Larson identified 18 independent pairwise differences or contrasts in the family Plethodontidae (Fig. 5.12). Higher nodes were reconstructed according to a parsimony procedure (Farris 1970; see Chapter 3). Difference scores were calculated for each of the 18 pairwise comparisons for three variables: *C*-value, and two measures of developmental rate (limb differentiation rate and limb growth rate). Estimates of the branch lengths in units of time were obtained from molecular data.

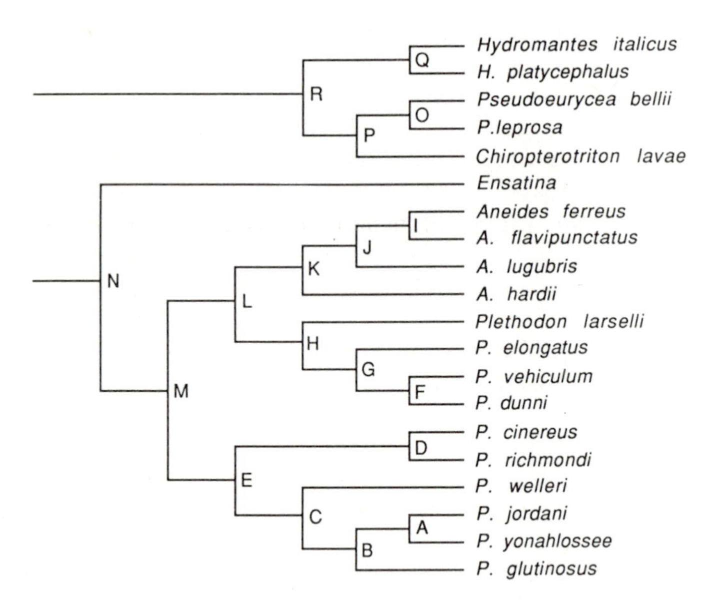

Fig. 5.12. The phylogenetic tree of the Plethodontid salamanders used by Sessions and Larson (1987) for their comparative analysis of limb differentiation rate and genome size. Letters assigned to nodes of the tree represent independent contrasts plotted in Fig. 5.13.

They tested their hypothesis for both variables by means of a rank correlation of the unstandardized pairwise comparisons. Limb differentiation rate but not growth rate was significantly inversely related to *C*-value. Large positive differences in the *C*-value within clades tended to go with large negative differences in the limb differentiation rate within clades ($r_s = -0.47$, $P < 0.025$; Fig. 5.13).

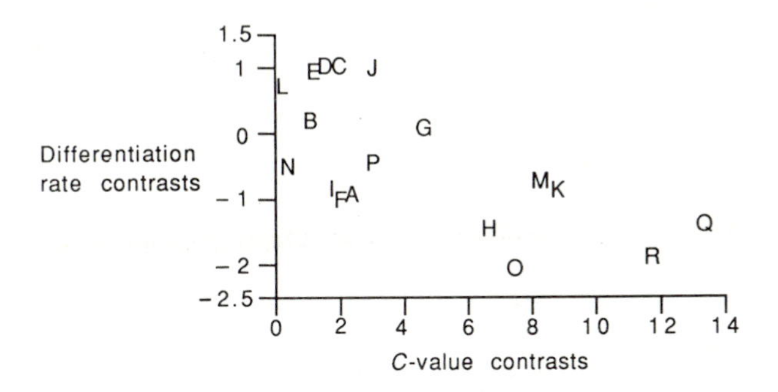

Fig. 5.13. Limb differentiation rate contrasts plotted against genome size (*C*-value, mass of DNA per haploid nucleus) contrasts for Plethodontid salamanders. The letters designate the nodes in Fig. 5.12. (After Sessions and Larson 1987).

Another application of Felsenstein's model is reported by Losos (1990), who studied locomotion and morphology in *Anolis* lizards. These lizards exhibit three locomotory patterns: running, jumping, and walking. Losos was interested in whether species that use one particular form of movement over another have morphological specializations for that behaviour. For example, species which typically run, such as sit and wait predators, may have evolved longer hind limbs.

Data on the percentage of movements attributable to running, jumping, and walking were collected from field observations of the 13 Jamaican and Puerto Rican *Anolis* species. Measures of fore limb, hind limb, and snout-vent length were obtained from 15 individuals from each species. Then, Losos used Felsenstein's method to analyse the relationship between morphology and the percentage of total movements that were walks. Twelve pairwise comparisons were calculated from the phylogeny in Fig. 5.14. Comparisons were standardized using branch length information obtained from literature sources. Hind limb length was negatively correlated with walking frequency controlling for body size (walking frequency as a percentage of all moves, $r = -0.75$, $P < 0.01$).

To summarize this section, Felsenstein's (1985*a*) method finds a set of independent pairwise differences or contrasts, each of which is scaled by its

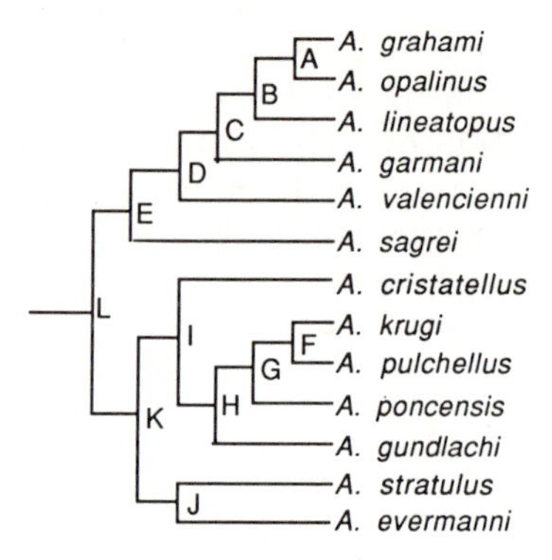

Fig. 5.14. Phylogenetic relationships among 13 species of Jamaican and Puerto Rican species of *Anolis* lizards. The 13 species allow 12 independent comparisons, labelled A to L, between the pairs of daughter taxa derived from each node. This phylogeny was used by Losos (1990) to examine whether evolution in limb morphology has been associated with evolution in locomotor propensities and movement (see also Section 5.10 and Fig. 5.22 in which this phylogeny is used to illustrate directional methods—each of the 12 nodes provides for two ancestor-descendant comparisons). Path lengths, not given here, were estimated by Losos using immunological and electrophoretic distance measures.

expected standard deviation. Expected standard deviations are derived by postulating an evolutionary model (Brownian motion) that translates branch lengths into units of expected evolutionary change. This illustrates how the choice of a comparative method is a choice of the way evolution proceeds as well as a choice of a set of statistical procedures. If, for example, a gradualist model of evolution such as this is assumed, and evolution has been punctuational (either everywhere or in some branches), or has proceeded at different rates in different branches, then the accumulation of variance will not be properly estimated, and the variates will not be scaled as intended (see Section 5.10, and Martins and Garland 1991).

These comments are not a criticism of Felsenstein's method. His procedures provide a way, in principle, of scaling the data according to whatever evolutionary model is proposed. Indeed, Felsenstein's procedures even allow for different rates of change in different branches.

5.8.3 Two Felsenstein-like approaches

Methods developed by Grafen (1989) and by Pagel and Harvey (1989*b*) follow Felsenstein's (1985*a*) idea of finding a set of independent comparisons. However, these methods can be applied to imperfectly resolved phylogenies, such as might be the case if a taxonomy was used in place of phylogeny. Here we describe how the methods work on a simple bifurcating phylogeny. We take up the case of imperfect phylogenies in Section 5.8.6.

Grafen's method

Felsenstein's method assumes that branch lengths are known and uses a null-model of character change to derive expected variances of the observations. Grafen (1989) assigns branch lengths according to a counting rule, and assumes that the expected variance of change along a branch is proportional to its length (Brownian motion model). Branch lengths are found by first assigning to each higher node, one fewer than the the number of species below it in the tree. Thus, species receive a zero, the next higher node is assigned one fewer than the number of species in it, and so on. Then, branch lengths are calculated as the difference between successive nodes. For example, the branch lengths leading to two species sharing an immediate ancestor would be one. Alternatively, if branch lengths are known by some independent means they can be specified directly.

The initial branch lengths are then lengthened or compressed in response to a parameter ρ that is estimated from the data. A maximum likelihood procedure is used to estimate ρ which alters all path lengths in the tree by a positive power that can vary between 0 and 1. In its extremes (i.e. where ρ is close to 0 or to 1), this parameter alters the branch lengths so that the majority of the variation in the tree is placed either: (1) close to the species level making species relatively independent (that is, branches leading to species are lengthened) or (2) close to the root of the tree making higher nodes more independent. Alternatively, ρ may lie somewhere in between (a single value of ρ is estimated, thus this stretching or compression of the tree is the same for all variables regardless of their individual distributions of variance throughout the tree). Values of ρ near 1.0 assign greater variance to higher nodes by stretching the higher branches; values nearer to zero assign greater variance to lower nodes by stretching the lower branches.

Once the estimates of the variances of change for different branches are obtained, Grafen's procedure finds comparisons among pairs of species or nodes, for a bifurcating phylogeny, like Felsenstein's method. The comparisons are scaled according to their estimated branch lengths, where branch length is a measure of expected variance. The parameter ρ is used to remove any correlation between the magnitude of a standardized comparison and its estimated variance. Then, on the assumption that the Brownian motion model provides comparisons that have been properly scaled, relations among two or more variables can be studied by standard correlation and regression techniques. Grafen (1989) reports simulation studies in which, for characters generated by a Brownian motion model of change, his method yields valid Type I error rates and has good statistical power. Stone and Willmer (1989) used Grafen's method to examine whether body size and thermal regime are related to warm-up rates in bees.

Pagel and Harvey's method

The method that we have developed finds the same set of comparisons as Felsenstein's or Grafen's methods but differs from those methods in the way that it scales comparisons for the expected variance.

The method assumes that rates of evolutionary change are likely to vary in different portions of the tree, and even within branches of the tree, and thus initially sets all branch lengths equal. For a bifurcating tree all comparisons are between two species or subtaxa. The method assigns all of these comparisons (differences) an expected variance proportional to two times the fixed branch length (variance of a difference between two independent data points is just the sum of their variances). Although this is implicitly a punctuational view of evolution, we do not mean necessarily to advocate that view. Rather, the method explores the possibility that arbitrarily fixed branch lengths can nevertheless yield acceptable statistical properties. At a later stage, patterns in the data are used to assess whether the scaling has in fact produced those properties (see p. 151) This method of scaling the comparisons has yielded approximately normal distributions of residual errors about regression lines in two recent studies (Harvey *et al.* 1990; Trevelyan *et al.* 1990).

As with the previous two techniques, the set of scaled comparisons can be analysed by standard regression and correlation techniques on the assumption that the scaling procedures result in a set of comparisons with equal expected means and variances under the null hypothesis[13].

5.8.4 Limitations of procedures for scaling comparisons

Many comparative studies of continuous variables combine in the same analysis, measures as disparate as mass, timing, and counts. It is reasonable to expect separate rates and distributions of change in these variables, or different rates and distributions of change in different regions of the phylogeny. If we had such information, it could then be applied

[13] The method has been used to examine relationships between clutch size and body size in birds (Blackburn 1990); metabolic rate and life history characteristics in mammals and birds (Harvey *et al.* 1990; Trevelyan *et al.* 1990); the size of the brain, its component parts, and ecological differences in mammals and birds (Harvey and Krebs 1990; Harvey in press); bird song, mating system, and life history (Read 1989); geographic range and habitat use in mamals (Pagel *et al.* 1991); parasite burden and geographical range in birds (see Harvey *et al.* 1991); and recombination rates and age at first reproduction in mammals (Eldred in press).

flexibly in Felsenstein's model. The computer application of Grafen's method, for example, allows the user to specify branch lengths (in units of variance of evolutionary change). However, lacking such information, we should be aware that the comparisons may not have the statistical properties we desire.

Formally, we should not use parametric statistics to analyse the comparisons derived from any of the three methods described in the previous two sections unless we know that the assumptions of those statistics have been met. Because the set of comparisons can reasonably be regarded as independent, the critical remaining assumption is that they all have the same variance. The Brownian motion model implicit in the methods does not guarantee this: equality of variances depends upon whether the Brownian motion model provides an accurate description of evolutionary change. In practice, correlation and regression techniques are quite robust to violations of their assumptions. However, this should not be taken as license to use them uncritically. We describe two techniques in this section that can be used to increase the chances that our statistical tests are valid: analysis of residuals, and randomization procedures.

Analysis of residuals

The relevant information for an analysis of residuals is the distribution of residual errors around the regression line formed by the relationship of the Y comparisons to the X comparisons. The residuals are found as the difference between the observed value of the Y variable and its predicted value as determined by the regression line. The frequency distribution of residual errors should be normal, with approximately 95 per cent of the points within two standard deviations of the mean. If the comparisons have not been scaled properly, then some will have larger true variances than others. This will manifest itself as heterogeneity of the variance of the residuals about the regression line.

We will concentrate on one technique for the detection of heterogeneity of variance among the residuals. Let s^2 be the variance and s the standard deviation of the residual errors about the regression line. It is a property of regression that the mean of the residual errors is zero. The vector

$$Z_e = \frac{\mathbf{e}}{(s^2)^{1/2}} \tag{5.7}$$

is the vector of residual errors divided by the square root of the variance (the standard deviation) of residual errors. This transforms the residual errors to standard scores, that is, scores with a mean of zero and a standard deviation of one. Many regression procedures automatically calculate 'standardized residuals' according to eqn (5.7). If the residuals have all come from the same underlying normal distribution, then a histogram of the elements $\mathbf{Z}_e$ should be normally distributed with approximately 95 per cent of its observations falling in the interval –2 to +2. This is easily checked by means of a binomial test: the number of observations outside of the –2 to +2 range should not exceed that expected by chance under the binomial given that $P = 0.05$, where P is the probability of being < -2 or $> +2$. Other more sophisticated tests can be used to assess the shape of the distribution.

The residuals also should show no tendency to change systematically with the predicted value of Y obtained from the regression equation. The correlation of the residuals with the predicted Y is always exactly equal to zero. Nevertheless, the spread of residuals might increase (or decrease) with the predicted Y, or the residuals may show a curvelinear pattern against the predicted Y, and still have a zero correlation overall. Either of these patterns suggests heterogeneity of variance.

In the face of significant heterogeneity of variance what can be done? Standard methods are available on most statistical packages (SPSS, Minitab, SAS, GLIM, and BMDP all have methods for treating heterogeneity of variance). Heterogeneity of variance means that some form of weighted regression is required. Alternatively, one might attempt transformations of the raw data or of the comparisons as a way of removing the heterogeneity of variance. And here we come full circle to the methods for scaling comparisons that have been described previously in conjunction with the various methods. This is because weighted regressions, and (non-linear) transformations work, in effect, by stretching or compressing individual data points as a way of equalizing residual errors. A very large positive residual error may suggest that the data point needs to be scaled downward, a very large negative residual may suggest the opposite. The only real difference in doing it at this stage is that the weighting is conditioned on patterns in the data, rather than in response to an assumed model.

Randomization tests

In this section we are concerned not with heterogeneity but with the problem of not knowing what the null hypothesis distribution is. Randomization tests provide a way to estimate the null hypothesis sampling distribution from the data (Bradley 1968; Sokal and Rohlf 1981). Then, the result observed in the raw data can be compared against the simulated null-model distribution of outcomes to obtain valid statistical hypothesis tests (examples from comparative biology are given in Pickering 1980; Harvey 1986; Elgar and Harvey 1987; Pagel and Harvey 1988*a*; Blackburn *et al.* 1990).

The general procedure of a randomization test is repeatedly to shuffle a data set and calculate some summary statistic each time as a way of generating a frequency distribution of outcomes under a given null hypothesis. For example, the null hypothesis distribution for a simple correlation between pairs of independent comparisons on *Y* and *X* might be simulated by randomly reshuffling the *Y* comparisons or the *X* comparisons and calculating the correlation. The histogram of correlations obtained from doing this a large number of times becomes the null-hypothesis distribution. The actual obtained correlation is checked against the empirically derived distribution of correlations to see if it is sufficiently large to consider that it is not a chance result.

Elgar and Harvey (1987) used randomization tests to analyse data on the relationship between basal metabolic rate and diet. McNab (1986*a*, *b*) had argued that basal metabolic rate in mammals was associated with diet even after adjusting for body size. Elgar and Harvey's objection was that diet categories are not evenly distributed among mammalian taxa, and thus differences in basal metabolic rate might be associated with differences among taxonomic groups. The relationship between diet and metabolic rate must be shown to hold independently of taxonomic association. Elgar and Harvey (1987) employed a randomization procedure that shuffled metabolic rates (adjusted for body size) and diet categories among taxonomic groups. This unconfounded taxonomy from the other two variables. After each of 2000 shufflings, they calculated the relationship between metabolic rate and diet. They then compared the actual empirical result with the distribution of results from their randomizations. They were able to confirm McNab's claim for only two of the 10 diet categories.

Sessions and Larson's (1987) study of the relationship of genome size and limb differentiation rate (Fig. 5.13) can be used to provide a useful illustration of randomization procedures. We re-analysed their set of 18 comparisons by a randomization procedure designed to capture the null-hypothesis distribution of two types of correlation coefficient: one is the regular Pearson correlation (r), and the other is the coefficient of congruence (r^*). The coefficient of congruence is not sensitive to the direction of change within a comparison (Harman 1967). For example, it does not distinguish between a pair of contrasts that are both positive versus a pair that are both negative. This is potentially important because the sign of a contrast is arbitrary: we have no basis for deciding between whether to subtract A from B versus B from A, where A and B are two daughter taxa or species.

Our procedure first randomly re-ordered the set of limb differentiation comparisons against the set of *C*-value comparisons, then calculated the two correlations on the randomized data. This was repeated 2000 times and the frequency distribution of results obtained. Figure 5.15 displays the frequency distributions for the two types of correlation. As would be expected, the frequency distribution for the Pearson correlation is centred roughly symmetrically around zero. The r^* distribution, however, has a smaller variance and is centred around –0.45. The slight skewing of the distributions is probably due to the distributional quirks of the 18 pairs of comparisons, and supports Sessions and Larson's expressed concern about testing the Pearson correlation against the tabled null-distribution values. However, the randomizations include these quirks in the data, and thus the appropriate probability values can be read right off the distributions. The obtained Pearson correlation of –0.65 is significant at the $P = 0.005$ (two-tailed) level. The r^* coefficient of –0.66 is also highly significant. These results agree with, but are slightly more extreme than, those reported by Sessions and Larson (1987).

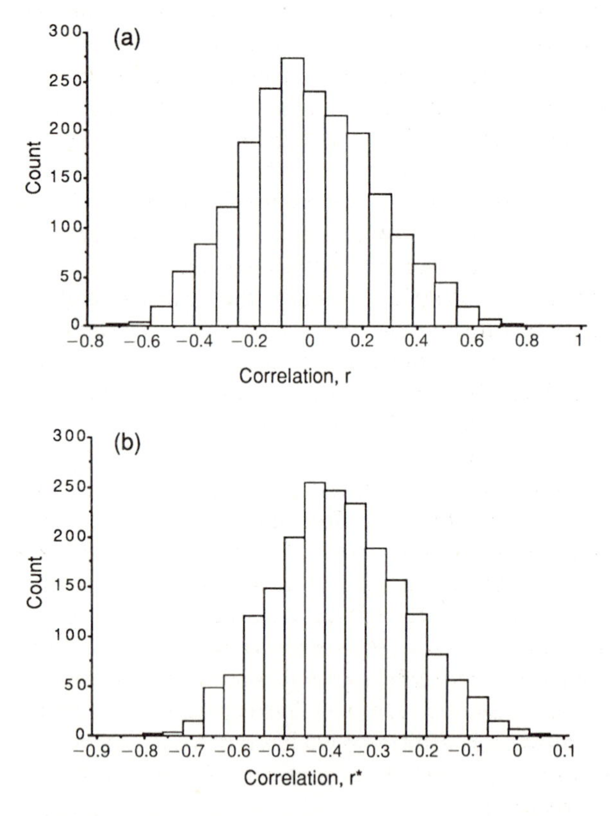

Fig. 5.15. The data from Fig. 5.13 have been randomized so that each differentiation rate contrast is paired with a random *C*-value contrast. The procedure was repeated 2000 times. On each occasion ('Count') a Pearson's product moment correlation (r) was calculated (a, above), as was a coefficient of congruence (r^*) (b, above). The coefficient of congruence, as described in the text, is not sensitive to the direction of change within a comparison.

Other procedures for using the data to derive the null distribution exist and are slowly receiving more notice. The bootstrap (e.g. Efron and Gong 1983) is probably the best known of these re-sampling methods and is based on more rigourous statistical theory than simple randomization procedures. Lunneborg (1985) and Wu (1986) describe applications of the bootstrap to testing the correlation coefficient.

5.8.5 Summary of independent-comparisons methods

Of the procedures we have described for conducting analyses separately within groups, three (Felsenstein 1985*a*; Grafen 1989; Pagel and Harvey 1989*b*) extract from the data set a series of phylogenetically-defined independent comparisons which each bear on the comparative hypothesis. The comparative relationship is assessed via the number of times it has independently evolved rather than by the number of species or higher nodes that have come to inherit it. The three procedures will extract the same set of comparisons from a bifurcating phylogeny, but differ in how those comparisons are scaled. The illustrative example in Box 5.2 ignores the problem of scaling.

5.8.6 Comparisons on incompletely resolved phylogenies

The methods described in Sections 5.8.2 and 5.8.3 require that the true branching phylogeny is known. But, it is often the case that we are ignorant of the true (probably) bifurcating phylogeny. Good phylogenies based on molecular techniques are becoming available, but they often resolve the phylogeny only to the level of subfamilies or tribes (e.g. Sibley and Ahlquist 1985), and they may even lack resolution at these levels (see Sarich *et al.* 1989).

In many cases we will have either a poor phylogeny or even a taxonomy to represent the branching of species. These incompletely resolved phylogenies typically will contain many multiple-nodes, that is, nodes from which more than two daughter taxa are represented as direct descendants. With more than two tips in a node the logic of finding a simple difference breaks down. We need methods to cope with multiple nodes if we are to apply the logic of comparisons developed above. Our assumption is that multiple nodes are monophyletic but do not actually represent true simultaneous radiations. Rather we assume that multiple nodes conceal some unknown branching pattern (Grafen 1989; Maddison 1989).

If the multiple node is not a simultaneous radiation then the tips of the node will not be independent. For example, a multiple node with three species may actually conceal the phylogeny of Figure 5.16. Species 2 and 3 share a phylogenetic history that species 1 does not. Not knowing this we can nevertheless make the assumption that the multiple node conceals at least one evolved difference: for example, the difference between species 1 and the node from which species 2 and 3 descended. Ignoring branch lengths, this difference can be represented as species 1−(species 2+ species 3)/2. This weighted difference score reduces the information in the three species points to a single point by subtracting the mean of species 2 and 3 from species 1. This represents our assumption that there is at least one evolved difference within the multiple node.

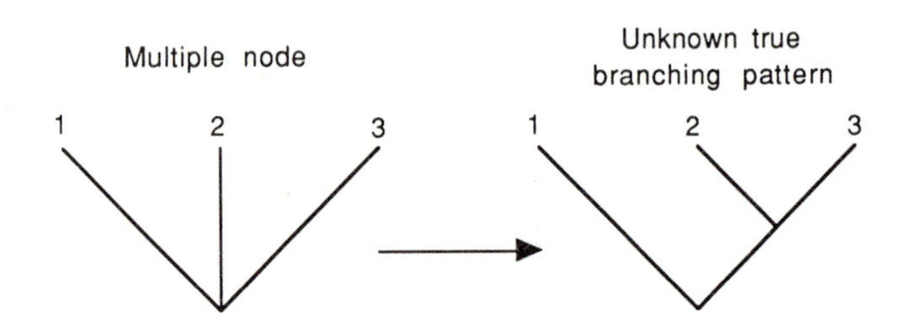

Fig. 5.16. A multiple node concealing an unknown true branching pattern. Taxa 2 and 3 share phylogenetic history that is not shared by taxon 1.

The example just given involves multiplying each species' value by a weight, and then finding the sum of the weighted values. If the weights are denoted by c_i, and the values of the character in each species denoted X_i, then the weighted sum can be written as $\Sigma c_i X_i$. Weights that have the property that they always sum to zero are known as *contrast coefficients*. Because the coefficients sum to zero, the weighted sum can be thought of as a weighted *difference* score. This logic can be applied to any number of points. The difference between two data points is the simplest case of a linear contrast, where the coefficients are equal to +1 and −1. In this case, then, the linear contrast is identical to what we would get by finding differences throughout a bifurcating phylogeny. Thus, we can think of simple difference scores as a special case of the more general problem of finding linear contrasts.

One way to view the contrasts coefficients is that they represent a hypothesis about the branching pattern of the unknown phylogeny. All of the tips that receive a positive weight are implicitly being treated as more closely related to each other than to the tips that receive a negative weight. The problem is that there are an infinite number of different linear contrasts for any given set of data points. Thus, we need to justify the methods for finding the contrast coefficients. We will describe the methods Grafen (1989) and Pagel and Harvey (1989*b*) chose.

Grafen (1989) chose contrast coefficients for multiple nodes by a procedure that gives greater weight to those species or nodes whose data points are not well explained by the phylogeny or by the other predictor variables under consideration. Grafen's procedure in effect first regresses the *Y* variable on to a series of dummy codes representing phylogenetic membership, plus any other control variables chosen by the investigator. The residuals from the regression of the *Y* variable will have the property that they sum to zero within each taxonomically or phylogenetically defined group (see for example the residuals in Box 5.1 for Stearns' method). The residuals are then used as the weights in the linear contrasts on the original *Y* and *X* variables, and a contrast is found for each presumed monophyletic group. Values of *Y* that are not well explained by phylogeny and by the *X* variables used as controls will have larger residuals, and thus be weighted more heavily in the contrast. A phylogenetic interpretation of this is that observations that deviate from the regression line in the same direction are more closely related to each other than those that deviate in opposite directions.

The method that we (Pagel and Harvey 1989*b*; Pagel unpublished manuscript) use to derive the linear contrasts relies on the assumption that the *X* (or *Y*) variable can provide useful information about the hidden phylogenetic structure in the multiple node. Multiple nodes are divided into two sub-nodes according to the distribution of *X*; those above the mean on *X* within the multiple node are assigned to group 1, those below the mean on *X* within the node are assigned to group 2. This criterion assumes that phenotypically more similar species (or higher taxa) within a multiple node are phylogenetically more closely related. If the *X* variable does not provide useful phylogenetic information, the comparisons may lack efficiency.

All branches within a node are assumed to be the same length. The contrast coefficients are found as the reciprocal of the number of species or groups within each sub-node. Then, the coefficients for group 2 are given a negative sign. Thus, for example, if the phenotypic criterion assigned three species to group 1 and two species to group 2, the contrast coefficients would be 1/3, 1/3, 1/3, −1/2, −1/2. This procedure is a slight modification of that which was used in an earlier version of the method. Contrast coefficients assigned this way always sum to zero. The method gives greater weight to the subgroup with the fewer number of species. In the case of a bifurcation, the coefficients will always be 1, −1. The set of coefficients is applied to *X*, to *Y*, and to each of the control variables to find a weighted difference score on each. The value of each weighted sum is standardized by dividing by the sum of the absolute values of the contrast coefficients. This procedure is repeated for each multiple node. Pagel (unpublished manuscript) describes this procedure as well as a more general procedure for taking into account either known or estimated branch lengths.

The set of weighted 'difference' scores derived from the separate independent comparisons can be used to estimate not only whether *Y* and

X are related, but to estimate the nature of that relationship as well. As we shall see in Chapter 6, the slope of the relationship derived from the analysis of the linear contrasts across taxa will estimate the slope of the relationship between Y and X, provided that the contrast coefficients are found in such a way that they are uncorrelated with the residual variation in the Y variable. Estimating the slope from independent contrasts avoids the problems of using non-independent species values (see Sections 5.3 and 6.6.2).

5.8.7 An example of independent-comparisons with unresolved phylogenies

Earlier in this chapter, for illustrative purposes, we used Millar and Zammuto's (1983) data set to examine the relationship between age at maturity and life expectation among mammals. When we employed the higher nodes method (Section 5.4) there was a significant positive relationship between the two variables independently of body weight. But when we used Stearns' phylogenetic-subtraction method, the relationship was not significant (Section 5.5). We pointed out that each method discards the information used by the other, and that independent comparisons allowed the use of all available information. Is the relationship significant when independent-comparisons methods are used?

To develop the example, we have continued to use the simple taxonomy of species, genera, families and orders as in Section 5.5, and have calculated the linear contrasts and independent comparisons for body weight, age at maturity, and life expectation at maturity. The 29 species classified according to the standard mammalian taxonomy used in Section 5.5 (Corbet and Hill 1980) allowed the calculation of 17 independent comparisons for each variable. When size-independent life expectation comparisons are correlated with size-independent age at maturity comparisons, the relationship is highly significant ($r = 0.57$, $n = 17$, $P = 0.02$).

However, mammalian taxonomies provide only crude approximations of phylogenetic relationships. Using cladistic techniques on morphological data together with analyses of genetic variation recognised at the molecular level (see e.g. Benton 1988a, b), it is now possible to produce more accurate mammalian phylogenies which can be used with the independent comparisons methods. Using Millar and Zammuto's (1983) data set together with the most recent phylogenetic reconstructions drawn from many sources, we could distinguish 23 independent contrasts. With body weight effects controlled for, the correlation between life expectation at maturity and age at maturity is again highly significant ($r = 0.80$, $n = 23$, $P < 0.001$; Fig. 5.17).

The method can be used to reveal aberrant taxa or, as in this case, those most highly responsible for the relationship. The point at the bottom left of

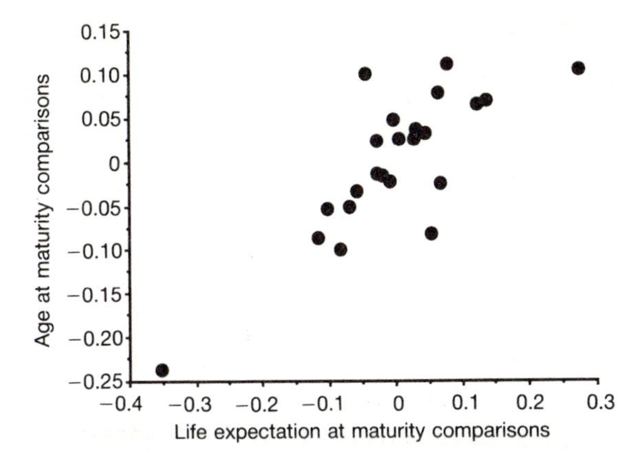

Fig. 5.17. Independent comparisons calculated from Millar and Zammuto's (1983) data. Age at maturity and life expectation comparisons are both corrected for body size in the figure, being given as a residuals from the regressions of each comparison on body weight comparisons. The correlation is highly significant (r = 0.80, n = 23, P <0.001). The bottom left point is a comparison between *Ochotona princeps* and *Sylvilagus floridanus*, and when this outlier is removed the correlation remains highly significant (r = 0.66, *n* = 22, P < 0.001).

Fig. 5.17 seems to be an outlier and represents a comparison between two lagomorphs, the northern pika *Ochotona princeps* and the Eastern cottontail *Sylvilagus floridanus*. The significance of the correlation does not depend on that comparison (with the comparison removed $r = 0.66$, $n = 22$, $P < 0.001$).

5.9 Testing hypotheses with independent comparisons

Under most circumstances, testing hypotheses with independent comparisons on Y and X proceeds as in Figs 5.17 and 5.18(a). When the relationship between the Y and X comparisons is positive and all or nearly all comparisons are positive, a simple linear regression or Pearson correlation (or non-parametric equivalent) can be used. If the X comparisons are positive and the Y comparisons are negative, Fig. 5.18(b), the same procedure can be used.

However, some patterns in the Y and X comparisons require more care in their interpretation. Specifically, it is necessary to test whether the slope *and the intercept* of the regression of the Y comparisons on the X comparisons differ from zero. Figure 5.18(c) depicts the case where all comparisons on Y and X are positive, but the magnitude of the Y comparisons does not change with changes in the magnitude of the X

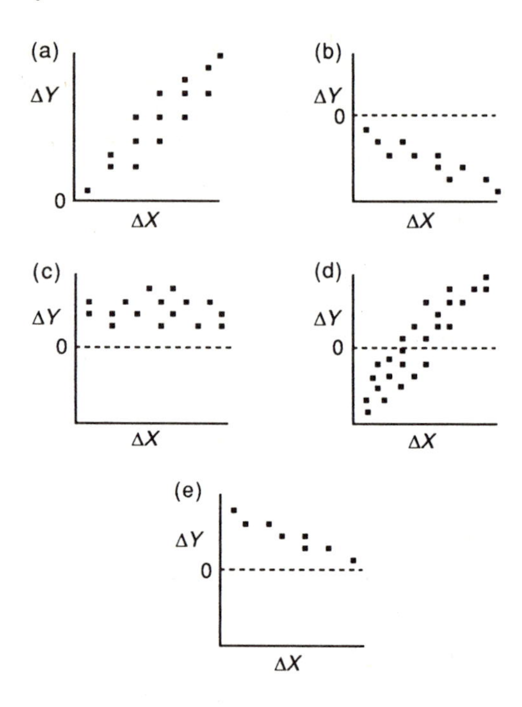

Fig. 5.18. Hypothetical patterns of relationships between sets of independent comparisons in which the comparisons on X (ΔX) are constrained to be positive. Comparisons on Y (ΔY) can be either positive or negative, and can be positively or negatively correlated with comparisons on X.

comparisons (slope zero, intercept greater than zero). This pattern, nevertheless, is strong evidence against the null hypothesis of no relationship between changes in Y and X, which would predict an equal number of negative and positive comparisons on Y. A standard linear regression on these data would find no relationship. However, either a regression forced through the origin (Grafen 1989) or a simple binomial test will detect a significant relationship. If the former is used, however, the slope of the line should not be interpreted, only the sign of the slope is of interest. Figure 5.18(d) shows a relationship that appears significantly positive, but on reflection reveals that the relationship between Y and X goes in the positive direction in about half of the taxa and in the negative direction in the other half (slope and intercept differ from zero). A regression through the origin or a binomial test would correctly indicate that no significant relationship existed. Figure 5.18(e) shows an instance in which a standard linear regression would find a significant negative relationship (again, slope and intercept differ from zero). But, as all Y

comparisons are positive, it would be incorrect to interpret such a slope. Again, either a regression through the origin or a binomial test would treat this situation properly.

Many other patterns are possible and can be dealt with by applying the logic of one of the examples given here, but what realistic examples might give rise to patterns such as those depicted in Figs 5.18(c)-(e)? Figure 5.19 depicts a possible scenario.

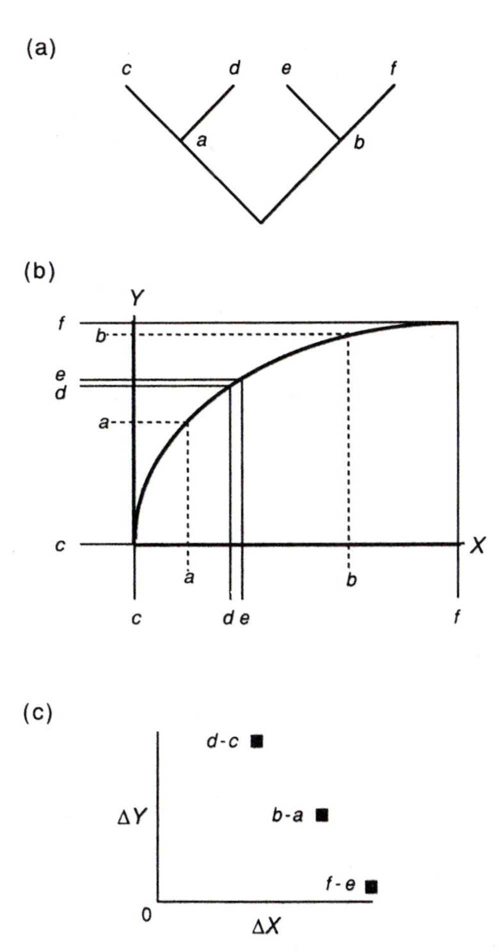

Fig. 5.19. The phylogeny in (a) shows four extant species and two ancestral nodes that can be used to make comparisons. *Y* and *X* are related by the curve shown in (b), with the values for nodal taxa *a* and *b*, and the four extant species *c* to *f* being depicted on the figure. Independent comparisons on *X* and *Y* (ΔX and ΔY) are shown on (c), and are negatively correlated even though a phylogenetic increase in *X* is associated with an increase in *Y* for each comparison.

Figure 5.19(a) shows a phylogenetic tree, Fig. 5.19(b) shows a non-linear relationship between Y and X in the raw data for the taxa in that phylogeny, and Fig. 5.19(c) shows a plot of comparisons similar to that in Fig. 5.18(e). The points a to f on the X and Y axes in Fig. 5.19(b) indicate the hypothetical values for the four species and the two higher nodes. Values of $b-a$, $d-c$, and $f-e$ form the three independent comparisons from the phylogeny. The relationship between Y and X in Fig. 5.19(b) is such that the small difference between d and c on the X axis (indicated by the vertical lines drawn up from those points) translates into a large difference on the Y axis. The large difference between e and f, on the other hand, translates to a small difference in Y, and the difference between b and a is intermediate on both axes. Note that the three changes in X are correlated with their average values of X. By appropriate non-linear transformation of the Y axis, this example can generate either of the relationships shown in Figs 5.18(a) and (c). Inverting the Y axis allows other relationships, including that in Fig. 5.18(b), to be generated. In practice, it is often possible to transform Y and X so that the relationship between them is linear and the difference between the values for taxa being compared is independent of their average value on the X axis (see Chapter 6). When such transformations are made, regression intercepts tend not to differ from zero and difficulties of interpretation are minimized.

5.10 Directional methods

The methods we have described so far do not test the direction of change in two or more variables along the branches of a phylogeny. Thus, we might have two species that differ in Y and X in the same direction, but this alone does not tell us the direction of change from ancestor to descendant. Both species could have evolved larger values of X and Y than their immediate ancestor, they could have both become smaller, or even evolved in opposite directions. We reported a similar distinction in Chapter 4 between Ridley's (1983*a*) and Maddison's (1990) methods. Huey (1987) points out that directional methods are useful for studying rates of change over evolutionary time, for assessing the ecological conditions that may have selected for derived traits, and for analysing whether changes in two traits are coincident. Chapter 1 discussed further distinctions between directional and non-directional comparative methods.

Huey and Bennett (1986, 1987) report a method for assessing directional changes in continuous characters along the branches of a phylogeny. They studied directional and non-directional trends (see Chapter 1) relating preferred body temperature and optimal temperatures for running speeds in Australian scincid lizards. Preferred body temperature is defined as that temperature selected by the animal when exposed to a thermal gradient.

Huey and Bennett were interested in whether the temperature at which a lizard species runs fastest has evolved to keep pace with preferred body temperatures.

To assess the direction and rate of evolutionary change in these two variables, Huey and Bennett first had to reconstruct the ancestral states of the lizard phylogeny. They used an iterative procedure (suggested by J. Felsenstein) that estimated each higher node as the average of its nearest three neighbours, subject to the provision that the final set of higher nodes minimized the squared change in the links of the tree, summed across all links (see Chapter 3). W. Maddison (personal communication) has shown that, if all branch lengths are assumed to be equal, this procedure also yields the maximum likelihood set of changes under a Brownian motion model. The phylogeny that Huey and Bennett used and the reconstructed ancestral character states for preferred body temperature are shown in Fig. 5.20.

Fig. 5.20. A tentative phylogenetic tree of seven Australian skink genera. At the top left, the non-Australian *Mabuya* is used as an outgroup. Australian genera, listed from left to right are *Egernia, Tiliqua, Leiolopisma, Eremiascincus, Sphenomorphus, Hemiergis,* and *Ctenotus*. Numbers at tips are generic averages for thermal preferences (^{0}C). Numbers at nodes are presumed ancestral preferences generated by a minimum evolution method. (After Huey and Bennett 1987).

The directional relationship between optimal running temperature and preferred body temperature was assessed by calculating the regression of changes in optimal running temperature against changes in preferred body temperature. Changes were calculated as the difference between generic means and the nearest higher node. The authors chose to use genera because of concerns that species were not independent, although they acknowledge that the same problem, if not as extreme, may also apply to genera. Huey and Bennett (1986, 1987) did not adopt a model of evolutionary change from which they could derive the expected variances of change along the branches. Instead, the changes along the branches were treated statistically as if they were drawn from the same null hypothesis distribution, an explicitly punctuational model of evolution. The regression of changes in optimal running temperature on to changes in preferred body temperature yielded a slope of 0.25, significantly less than 1.0, indicating that directional changes in optimal running temperature have lagged behind changes in preferred body temperature (Fig. 5.21).

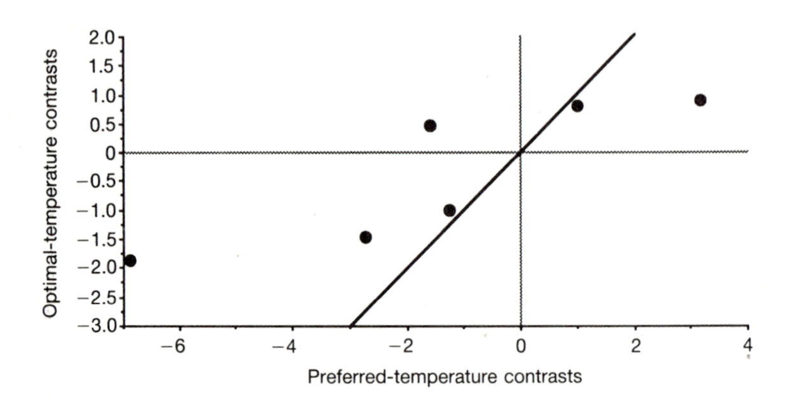

Fig. 5.21. Optimal running temperature versus preferred body temperature comparisons in Australian skinks. Change is measured as the difference between each of six generic values and its nearest node. Adaptation is seen to be partial because the change in optimal running temperature is less than that in preferred body temperature. The line represents perfect coadaptation where changes in optimal and preferred temperatures are the same. (After Huey and Bennett 1987).

Losos (1990) employed Huey and Bennett's procedure to study directional changes in locomotory behaviour and morphology in *Anolis* lizards (see Section 5.8.2 for a description of this study). Figure 5.22 plots the directional changes in locomotor behaviour versus directional changes in hind limb length, controlling for body size.

Huey and Bennett's procedure differs from the independent-comparisons methods by examining the changes between ancestors and descendants rather than between contemporary species or higher nodes. This allows tests of explicitly phylogenetic hypotheses about the direction and rates of evolutionary change that are not directly available to non-directional methods. In practice with this method, as the authors note, there will be some dependence among the lower and higher nodes because the higher nodes are estimated from the lower nodes. For example, in the extreme case, we might estimate the ancestral state of a genus as the arithmetic mean of the species. Because the sum of the deviations of the

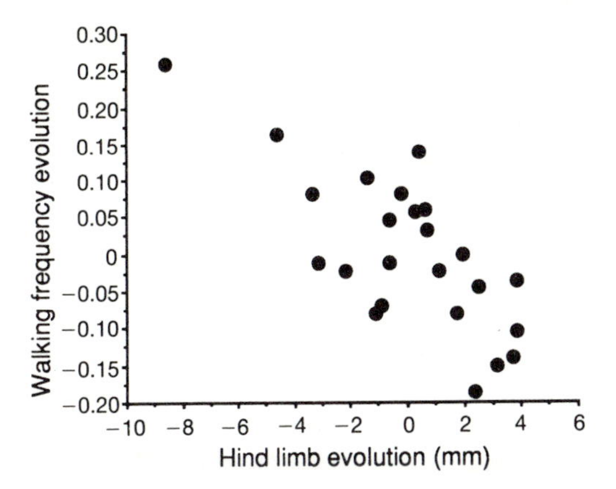

Fig. 5.22. The relationship between the evolution of hind limb length and walking frequency in *Anolis* lizards using Huey and Bennett's (1987) method. Walking frequency is walking as a proportion of all moves, and hind limb evolution is the residual of the change in hind limb length relative to the corresponding change in snout-vent length. Each point corresponds to the amount of evolutionary change in one ancestor-descendant pair. (After Losos 1990).

species about their mean must sum to zero, for a genus with k species only the first $k-1$ species in the genus can be independent. If changes along branches are computed as deviations from higher nodes, the property that the deviations sum to zero forces some species to appear to have increased and others to have decreased with respect to the ancestral condition. However, the effects of this non-independence do not appear to be critical (see Section 5.11).

5.10.1 Accumulation of variance over time

Bell (1989) reports a novel use of the nested analyses of variance to study the accumulation across taxa of diversity in characters over time. This technique is useful for illustrating the rate of evolutionary diversification with time, and for identifying time periods of rapid evolutionary change.

As part of a larger study, Bell collected information on body weight and chromosome number in mammals. A nested analysis of variance was conducted on each character using seven taxonomic levels. Bell also collected information on the approximate times of divergence of the taxonomic levels, and then plotted the cumulative percentage of variance at each taxonomic level against time of divergence. The results are shown in Fig. 5.23.

If Bell's analysis is based on correct assumptions, approximately 19 per cent of the variance in body mass that was eventually to appear among contemporary species was already present among super-orders within the

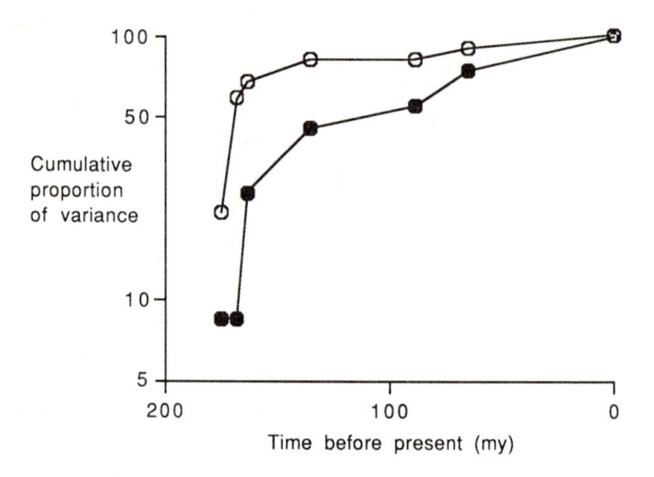

Fig. 5.23. The increase in variance of body mass (○) and chromosome number (●) through time in eutherian mammals. The eight taxonomic levels used for the analysis to produce seven nested variance components are infra-class, superorder, order, suborder, family, subfamily, genus, and species. (After Bell 1989).

infra-class Eutheria during early Tertiary times. Diversity in body mass among taxonomic groups increased sharply through the radiations of suborders by which time 78 per cent of the variance seen today was already present. The rate of diversification in body mass fell off sharply afterwards. Diversification in chromosome number lagged behind that for body mass at all stages. This means that the variation among taxonomic groups in chromosome number was not as great as that for body mass. One interpretation is that the variation in chromosome number within groups is always somewhat large compared to that across groups. Recalling the interpretation of the cumulative percentage of variance as an intra-class correlation coefficient (Section 5.4), the chromosome number of one member of a group will not be a very good a predictor of the chromosome number of another member of the same group. Bell (1989) suggests that chromosome number may have been more often associated with speciation events than was body size.

Figure 5.23 suggests that the amount of divergence among species is closely related to time since they shared a common ancestor. Bell tested this idea by plotting the variance of body mass among species within a genus against the age of the genus for 64 different genera. The plot, which showed a highly significant positive relationship ($P = 0.002$), confirms this idea.

5.11 Selected computer simulation results

We finish this chapter by presenting selected results from a series of computer simulations conducted by Martins and Garland (1991) to compare the performance of each of several different comparative methods under different evolutionary scenarios.

Our interest in Martins and Garland's work is not to use it to argue for the superiority of one technique over another. Such a conclusion would depend upon the models of evolution that were used to generate the simulated data sets being representative of the processes responsible for the diversity in real data sets. Rather, we use their results to emphasize that the success of a method depends strongly on whether the data being analysed were generated by the evolutionary model on which the method is based.

Martins and Garland analyzed the performance of five different comparative methods. The methods included a simple cross-species regression, Felsenstein's (1985*a*) method with standardized and unstandardized comparisons (i.e. the latter are not divided by an estimate of their standard deviation), and the method reported by Huey and Bennett (1987), again with standardized and with unstandardized changes. The standardized methods based on Felsenstein's approach were found according to two rules. The FL1G procedure standardized according to a Brownian motion gradual model of evolution in which variance accumulates additively over time. The FL1P, or punctuational, version allowed only one unit of variance per branch (i.e. all branches the same length). The non-standardized versions FL2G and FL2P are identical to the FL1 versions except they are not standardized.

Huey and Bennett's procedure was also represented by four versions. ME1G (for minimum evolution) calculates changes between all connected points on the phylogeny. Higher nodes were calculated as weighted averages of lower nodes, taking into account branch lengths. ME1P is the same as ME1G except all branch lengths are assumed to be equal. ME2G and ME2P replicate the ME1 versions except changes are calculated only between nodes and tips of the tree (ME2P corresponds most closely to what Huey and Bennett actually used). Felsenstein's, and Huey and Bennett's methods were then simulated for both punctuational and gradual models of change.

In the first of their simulations Martins and Garland studied Type I error rates, α, from simulations of a 'known' phylogeny of 15 species (see lower portion of Fig. 5.12, after Sessions and Larson 1987) with varying branch lengths. Data were generated by a Brownian motion process, with no correlation between Y and X. Branch lengths were measured in units of expected variance. Thus, this data set conformed exactly to the model assumed by Felsenstein's method. Felsenstein's method yielded a Type I error rate very close to the nominal α of 0.05 (see Table 5.5) for this case. Interestingly, however, the other versions of Felsenstein's model had Type I error rates only about 3 to 4 per cent higher than the exact model, a difference which was significant statistically, but perhaps not very large in practice. The simple correlation across species had a very high Type I error rate. The minimum evolution models (after Huey and Bennett 1987), especially ME2G and ME2P, had somewhat higher Type I error rates, as might be expected given that changes along the branches are not independent (see Section 5.10).

Table 5.5 Type 1 error rates and power of several comparative methods to detect a false null hypothesis. For details of analysis models, see text. 'G' and 'P' refer to gradual and punctuational change respectively. 'Species' denotes a cross species regression. (After Martins and Garland 1991).

	Type 1 error rate		Statistical power			
Analysis model	Simulated G change	Simulated P change	Simulated gradual change for different values of ρ			
	α=0.05	0.05	ρ=0.25	0.50	0.75	0.90
FL1G	0.05	0.08	0.16	0.53	0.94	1.00
FL2G	0.07	0.06	0.13	0.40	0.87	1.00
FL1P	0.08	0.05	0.14	0.42	0.89	1.00
FL2P	0.09	0.06	0.13	0.39	0.86	1.00
ME1G	0.09	0.06	0.12	0.36	0.85	1.00
ME2G	0.14	0.09	0.10	0.28	0.75	0.98
ME1P	0.07	0.05	0.16	0.44	0.90	1.00
ME2P	0.10	0.08	0.14	0.38	0.85	0.99
Species	0.16	0.16	0.08	0.26	0.68	0.97

In contrast, when Martins and Garland simulated data according to a punctuational model of change, a different set of results emerged. In this set of simulations, a single punctuational change was allowed in each branch, regardless of its length. Here, Felsenstein's gradual model of change had slightly elevated Type I error rates, about 2.5 per cent higher than the two punctuational models, which had Type I error rates very close to 0.05 (Table 5.5). The minimum evolution methods performed better with punctuational change than with gradual change but again, ME2G and ME2P did somewhat worse than the other minimum evolution models.

These results demonstrate a point that we have emphasized throughout this chapter: the success of a comparative method depends to a large extent on whether the assumptions that it makes about the way evolution proceeds are in fact true. However, it is important to point out that, aside from the species regressions, most of the Type I error rates, even for incorrect models, are not wildly elevated. Martins and Garland's simulations show that gradual and punctuational models provide reasonably approximate Type I error rates even when applied to data derived from the alternative model. However, in both cases the comparative method whose underlying model best matched that used to generate the data did significantly better, and not different from what would be expected theoretically.

Martins and Garland also conducted simulations to establish the relative power of the methods under punctuational and gradual models of change. For a range of simulated true correlations between Y and X, the punctuational and gradual versions of Felsenstein's method had similar power (Table 5.5). The minimum evolution methods generally had somewhat lower power.

5.12 Conclusions

We hope to have made clear throughout this chapter the inescapable connection between the statistical procedures that a comparative method employs to create a data set suitable for analysis by parametric techniques, and the evolutionary assumptions necessary to justify those procedures. As logic dictates, and simulation studies confirm, when a model's assumptions are true, the model performs well. When they are not met, the models do not perform as well. Nevertheless, models must be countenanced in spite of their weaknesses because, as we stated in the previous chapter, techniques which ignore them are liable to make implicit assumptions that are even less realistic than those made by the models. Comparative methods must either deal with assumptions directly in the form of explicit models which may be wrong, or acknowledge that they depend upon highly constrained evolutionary scenarios. Either way, we want to emphasise that the consequences must be dealt with statistically, and we have described methods of residual analysis to that end.

A niggling problem with independent-comparisons methods is that we do not as yet have a way of estimating ancestral characters that is independent of the distribution of extant characters. This can introduce some dependence among the members of a set of comparisons. Felsenstein

(1985*a*, p. 13), relaying an idea suggested by a student, suggested that 'we could use comparisons between pairs of species that we were fairly sure had a common ancestor not shared with any member of another pair, and that these contrasts could then be safely assumed to be independent'. Burt (1989) later reported the same technique, but without the evolutionary considerations with which Felsenstein imbued his model. This dependence does not seem to be a serious problem, however. If it were, we should not expect independent-comparisons methods to have performed so well in Martins and Garland's (1991) simulations.

On logical grounds, independent-comparisons methods and methods that use explicit ancestral character reconstruction to test the direction of evolutionary trends should be preferred over their cross-species rivals. The best developed cross-species methods unnecessarily discard large amounts of information that independent-comparisons and ancestral reconstruction (i.e. directional) methods approaches exploit. There is no good reason to discard as inappropriate for testing adaptive trends, the variance that is correlated with phylogenetic differences. We simply must know how to treat this variance, and independent-comparisons methods provide the way.

5.12 Summary

The branching structure of phylogenies ensures that species are not independent for statistical purposes. Various comparative methods differ in how they estimate and manage this non-independence. Some methods discard information in an attempt to create a set of independent points, while others which make use of all the variation in the phylogeny are to be preferred on logical grounds. These methods employ independent comparisons either to assess differences among species or higher nodes, or to assess the direction of evolutionary change. Evolutionary models implicitly or explicitly underpin all methods. The validity of a method depends upon whether the model on which it is based accurately describes the evolutionary processes that have generated diversity.

6

Determining the forms of comparative relationships

Allometry 'explains, among many other things, why large animals have relatively thick legs, why dachshunds can't be as large as elephants, [and] why flies can walk up walls' (Gould 1975, p. 245)

6.1 Introduction

Allometry has been variously defined as 'the study of size and its consequences' (Gould 1966, p. 587), 'the biology of scaling' (Calder 1984, p. 1), the study of 'the structural and functional consequences of changes in size or scale', (Schmidt-Nielsen 1984, p. 7), and perhaps more daringly as 'the study of the relationship between size and adaptation' (Fleagle 1985, p. 1). What is at issue in each of these quotations is the pervasive tendency for a variety of morphological, physiological, life history, and behavioural characteristics of organisms to show highly regular changes with changes in body size. Allometry is used to describe such relations quantitatively, and we use it here to illustrate a situation in which the comparative biologist's primary interest is not in whether a relationship between two variables exists but in the form that the relationship takes.

This chapter briefly reviews the field of allometry, and then moves on to describe how to estimate allometric relationships using the logic of independent comparisons that were described in Chapter 5. Later, we outline one approach to giving a theoretical basis to allometric slopes. Much of allometry is purely descriptive, and we suggest that an understanding of allometric relationships may be found in the details of the selective forces acting on the variables under study. We illustrate this idea with data on life history variation in birds and mammals. Finally, we sketch some of the difficulties that arise in estimating allometric relationships when the variable under study is caused by several factors.

6.2 The fundamental allometric relation

The simple statement that allometric relations describe changes in characters with changes in body weight does not imply a particular underlying mathematical model. Nevertheless, from the pioneering work of Sir Julian Huxley (see Huxley 1932) it became apparent that many empirical relationships involving size are well described by the power relationship:

$$Y = \alpha X^{\beta} \tag{6.1}$$

where X is body weight, Y is the character thought to vary with weight, and α and β are the parameters of the power equation. The implication of this relationship when $\beta \neq 1$ is that the ratio Y/X (e.g. organ weight/body weight) changes with body weight. For example, brain weight in smaller mammals may be around 5 per cent of body weight, whereas the brain is only about 0.05 per cent of body weight in the larger whales. Equally dramatic examples come from single species: among juvenile male fiddler crabs, the large claw is about 2 per cent of body weight, but among large adults the claw may weigh up to 70 per cent as much as the rest of the body (Harvey and Clutton-Brock 1983). If the power equation indeed represents a general biological scaling law, then organ size, timing of life histories, and magnitudes of behaviours, all change differentially with body weight when $\beta \neq 1$: bigger animals are not just scaled-up versions of smaller animals.

When logarithms are taken of both sides of the power equation it becomes linear so that:

$$\log(Y) = \log(\alpha) + \beta \log(X). \tag{6.2}$$

The logarithmic transformation also tends to equate the variances of Y for different values of the X variable (in the raw data Y is likely to be positively correlated with its variance for samples with a broad range of body weights). From the logarithmically transformed data, the exponent of the power equation is estimated as the slope of the line. This slope is a measure of the differential changes in Y given X. Slopes greater than and less than 1.0 correspond to *positive* and *negative* allometry, respectively, terms which probably do more to obfuscate than to clarify. *Negative* allometry (not to be confused with a negative slope) indicates that the ratio Y/X decreases as X gets larger. The vast majority of allometric slopes are less than 1.0. If the slope exceeds 1.0 (*positive* allometry) Y/X increases faster than X. Antler length in deer and home range area in mammals are examples of positive allometry. A special case is a slope of 1.0, which indicates *isometry* or proportional change in Y and X. The parameter α is the Y-axis intercept, and if two lines of the same slope are compared, the difference between their respective values of α indicates differences independent of size. Such differences can often be related to differences in environment or behaviour, as we saw with testes size in Chapter 1 and as we shall see with antler length in Section 6.4.2.

As several authors have pointed out with etymological delight, the term *allo*metry is derived from the Greek *allos* or other, and simply refers to relationships in which the slope of the line is different from 1.0. There is no necessity that the relationship between Y and X be best described by a power equation for it to be termed allometric, however. Simple linear relationships of the form $Y = \alpha + \beta X$ are allometric if α is different from zero. This can be seen from the fact that only for $\alpha = 0$ is the ratio Y/X a constant, and thus changes in X are associated with proportional changes in Y.

It is nonetheless true that many relations can be expected on theoretical grounds to follow a power function. Surface area of a three-dimensional object increases as the square of increases in linear dimensions, while volume increases as the cube of the linear dimension, so long as geometric similarity is preserved. Changes in body weight can be expected to be associated with changes in body surface area and volume that roughly follow power functions, then, unless the organism drastically alters its shape. These changes in turn have implications for a host of physiological and structural systems such as bone size, metabolism, and respiration (see Gould 1966; Peters 1983; Calder 1984; Schmidt-Nielsen 1984). For example, smaller mammals have the highest weight-specific metabolic rates, perhaps (in part at least) to offset their much higher surface-to-volume ratios. Thus, Calder (1984, p. 4) remarks 'if [an] elephant had a metabolic rate as intense as that of a mouse, it could not dissipate the heat fast enough to avoid being cooked'.

But, lacking a theoretical specification of the relationship between some character and body weight, there is no *a priori* reason why it must be described by a power function, and a scatter plot of the untransformed data should be inspected for evidence of a power relationship. Indeed, the use of the power equation is often an empirical convenience rather than a biological necessity. For example, the assumption that many individual body parts scale allometrically with size rules out the assumption that the sum of these parts scales allometrically, because the sum of two power functions is not a simple power function itself (although deviations from a power curve may be slight and well within the realm of measurement error). The assumption of a power relationship may be more tenuous for intraspecific than for interspecific data sets. Smith (1980) analysed 30 interspecific and 30 intraspecific data sets and found that logarithmic transformations significantly improved the correlation coefficient (used to assess the strength of the linear relationship) between the Y variable and body weight in only 12 cases. However, re-analysis of Smith's data shows that in 83 per cent of the interspecific but only 47 per cent of the intraspecific data sets logarithmic transformation increased the magnitude of the correlation coefficient (Harvey 1982). Nevertheless, despite these caveats about the assumption that Y and X are related by a power function, unless otherwise stated it is the logarithmic equation of allometry that we will have in mind as we discuss allometric relationships in this chapter.

6.3 Types of allometric relationship

It is crucial to distinguish among the several different types of allometric relations (Cheverud 1982). Perhaps the best known allometric relationship is 'Kleiber's law' which describes the tendency for metabolic rate to increase with the 0.75 power of body weight across species of mammals (Kleiber 1932, 1961). With the addition of a few more points to Kleiber's original 26 species, Brody *et al.* (1934) published what is now the classic mouse-to-elephant curve that replicated Kleiber's finding. Relationships such as these, fitted across average values for the adults of contemporary species, are known as *interspecific allometry*. Other well-known examples of interspecific allometry are the relationships between brain weight and body weight, and those relating mechanical characteristics of the skeleton to body weight, the latter dating back to Galileo. So common are interspecific relations with body weight that entire books have been devoted to them (McMahon and Bonner 1983; Peters 1983; Calder 1984; Schmidt-Nielsen 1984).

Some of the earliest work in allometry employed *intraspecific* scaling. As the name implies, intraspecific curves are fitted across individuals of a single species. The relationship is termed *static intraspecific allometry* when the individuals are all of the same age or stage of development. Lapicque's (1907) early work on the intraspecific relationship of brain weight and body weight in each of several mammal species touched off a controversy about the evolution of brain size and weight in mammals, which exemplifies what Gould (1975, p. 247) has labelled a 'problem of treatment' with respect to allometry. Lapicque found that among adults of a species brain weight tended to increase with about the 0.25 power of body weight, a finding that he proclaimed to be 'universal'. However, other work by Snell (1891) and Dubois (1898) had put the value at around 0.56 to 0.67 (now widely accepted to be between 0.67 and 0.75—see Harvey and Bennett 1983; Harvey and Pagel 1988). The fallacy of comparing Snell's and Dubois' work with Lapicque's is that the former's results were derived from interspecific rather than intraspecific studies. The relationship between brain weight and body weight within a species will be influenced by variation in body weight, perhaps due to environmental factors, that is independent of variation in brain weight; for example differences between obese and emaciated individuals. Thus, intraspecific and interspecific relations may describe different phenomena.

When allometry is used to describe the growth or development of a species, the intraspecific relationship is known as *ontogenetic* or *growth allometry*. Ontogenetic curves for brain and body weight are generally more complex than either static intraspecific or interspecific relations. During fetal growth the brain increases in weight in approximately direct proportion to increases in body weight. Later the rate of change of growth of the brain slows, and changes in body weight in adult life can be independent of changes in brain weight. The contradiction between this statement and our earlier statement that the static intraspecific allometry of brain and body weight had a value of about 0.25 may be due to the inclusion of immature individuals in many interspecific studies, or of males and females in species that are dimorphic in weight. Martin and Harvey (1985) showed that many static intraspecific relations were not different from zero when juveniles were excluded and sex differences in weight were controlled for.

6.4 Interpreting allometric lines

In the two previous sections we have outlined some of the technical interpretations of allometry, but this is not the whole story. How should allometric lines be interpreted biologically? Do they describe adaptive trends, or do they reflect constraints built in by weight? Can they be used to derive theory? Or, should they be used only to test theories derived on *a priori* grounds? The gulf of opinion on these issues is enormous. Peters (1983, p. 4), for example, contends that the collective data base of allometric equations comprise a theory of size, whereas Calder (1984, p. 2) states 'so far we lack a unifying theory of allometry'.

6.4.1 Allometry does not predict evolution

A common misunderstanding about allometric relationships is the belief that they describe the slope of the line of evolutionary change, given selection on body weight. For example, among adults of closely related mammal species the allometric slope linking brain weight to body weight is sometimes in the 0.20 to 0.40 range. As this falls within the range claimed for some intraspecific slopes (but see above), it has been assumed that 'ancestor and descendant may be the 'same' animal expressed at different sizes' (Gould 1975, p. 280). But Lande (1979) and Cheverud (1982) point out that this logic conflates static intraspecific allometry with what both authors call evolutionary allometry. The line of evolutionary change from one species to the next, given selection on weight, depends upon the genetic covariance between the Y variable and body weight, not the phenotypic covariance, which is partly determined by environmental variation.

6.4.2 Allometry as a line of constraint

The allometric line is frequently used to partition the total variation in the Y variable into two parts, one associated with body weight, and the other independent of body weight. Sometimes, the part associated with body weight (that described by the allometric line) is thought to represent a weight-determined constraint on evolution, an idea as old as allometry itself (Snell 1891). Alternatively, correlations with body weight are used to argue against the need to give adaptive explanations for characters on the grounds that they arise as a result of the passive consequences of selection for genes that influence body weight and which have pleiotropic effects on other body parts. We see no reason to accept either argument as a general rule. The first argument implies that even given selective pressures to change, the character would be held back, presumably by its relationship with other characters, particularly body weight. As far as we know, no such constraint that is unlikely to be broken over evolutionary time has yet been demonstrated. Constraints do exist, but without further evidence there is no reason to believe that the allometric line defines a line of constraint. Furthermore, there is plenty of evidence for adaptive changes in organ size away from expected trends, given selective pressures to do so (as represented by differing environmental conditions; for examples see Harvey and Clutton-Brock 1983). Gould (1966, p. 618) states 'there has been, in my opinion, mistaken emphasis on the non-adaptive nature of simple allometric trends'. In this context, the term 'constraint' probably only indicates our general ignorance of the selective pressures maintaining a character, or of the range of options open to organisms as they evolve different sizes. Allometric trends fitted across a range of species from very different environments may subtly encourage such thinking by averaging over the effects of many different selective pressures. Taking these arguments together, we might more profitably view the allometric line as a line that summarizes the many (mostly unknown) selective forces that are correlated with size.

What about the argument that the presence of an allometric relation obviates the need for adaptive explanations? Lewontin invoked this argument against the adaptive explanations offered for antler size in deer when he wrote that, due to the correlation of antler size with body size, 'it is then unnecessary to give a specific adaptive reason for the extremely large antlers of deer' (Lewontin 1979, p. 13). It may be possible to predict antler length reasonably well from body weight, but this should not preempt the search for adaptive explanations. Deer use antlers for fighting, and large antlers are found among the males of species with polygynous mating systems in which greater male-male competition can be expected. Clutton-Brock *et al.* (1980) were able to demonstrate that the males of deer species with polygynous mating systems had larger antlers for their body weight than those with monogamous or mildly polygynous mating systems.

The lesson to be learned from the deer's antlers example, and the many others like it (see Harvey and Clutton-Brock 1983), seems to be that the presence of strong allometric trends may often indicate the presence of strongly correlated selective pressures. The forces selecting for large antlers in deer may be the same as those selecting for large body weight in those same deer. When we observe strong allometric trends, then we should be just as curious about the selective forces responsible for them as we would be about the selective forces responsible for the variation in any one trait. Complex adaptations, such as large antlers, do not come free to organisms. They are costly to produce and maintain. If they were unnecessary there is good reason to expect that they would be lost over evolutionary time, since organisms without them would presumably derive some energetic or other ultimately reproductive benefits. On these grounds alone we are justified in suspecting that allometric relations are maintained by selective forces acting on both variables.

6.4.3 Allometry as a line of functional equivalence

Gould (1975, p. 261) recommends using the allometric line as a 'criterion of subtraction' to remove the effects of size. Then, species can be compared on the basis of relative character size (the difference between actual size of the character and the size expected on the basis of body weight) without fear of confounding species differences with differences in size. Does this use of the allometric line assume that points along it are functionally equivalent, in the sense of demarcating how a particular organ must change in size with changing body weight to perform the equivalent function? (We include in this sense of functional equivalence passive changes in one character due to selection on another character with which it covaries genetically.) And, if we are to use allometric lines as criteria of subtraction, how should those lines be determined?

The answer to the first question is yes, functional equivalence is assumed, although we usually have no grounds for believing the assumption. Thus, by using the allometric line as a criterion of subtraction we assume implicitly that the remainder represents that part of the Y variable that is over and above the part 'necessary' for a given body weight. Relative values, then, provide a way to study variation in characters that is independent of variation associated with body weight. This is an absolute necessity if we are to study adaptation. So long as variation in a character can be explained by its correlation with body weight, we cannot attribute the variation to other selective forces operating independently of body weight. But we should be aware of the weaknesses of this approach. Suppose that in polygynous deer species there is direct selection on antler length because deer with longer antlers win more fights and thus achieve more matings. Then, the correlation between degree of polygyny in deer and antler length could properly be called adaptive, regardless of correlated differences in body weight among the species. However, if the correlation between antler length and body weight was strong enough, a 'criterion of subtraction' approach might miss the adaptive changes in antler length. The slope of the empirically determined allometric line presumably would reflect the correlated changes in antler length due to changes in body weight (in the absence of direct selection on antler length) plus the extra increases in antler length due to direct selection. Then, relatively large antlers might not be found among more polygynous species because the variation would have been removed by subtraction. An appropriately designed experiment might help to untangle the body-weight correlated contributions of the two components contributing to variation in antler length. We shall return to a similar problem in Section 6.4.4.

In most instances we believe that the effect of empirically derived slopes will be to mask adaptations rather than create them. But, as our example shows, we should not place much faith in empirically derived slopes *per se*. Our answer to the question of how allometric lines should be determined, then, is unsatisfactory, but grudgingly realistic. Ideally, the way around the problems associated with empirically derived relations would be to derive the allometric lines from theory rather than from the data. Unfortunately, this would bring the study of adaptation almost to a snail's pace as the selective forces underlying allometric relations are so poorly understood.

6.4.4 Using allometric lines to deduce other allometric relations

Can we deduce the form of the power relationship between two characters if each has a strong allometric relationship with body weight? The practice is widespread (see Peters 1983; Calder 1984; Schmidt-Nielsen 1984 for examples), but as we will show, the statistical nature of allometric relations makes this practice unwise, other than as the first step of a more comprehensive analysis.

Martin (1981) developed a theory linking neonatal brain weight in mammals to maternal metabolic rate. Allometric analyses seemed to show that metabolic rate and adult brain weight were each related to adult body weight raised to the 0.75 power, implying that adult brain weight and metabolic rate may be related isometrically. Other analyses had shown that neonatal and adult brain weight were related isometrically. Martin (1981) deduced that metabolic rate and neonatal brain weight would also be isometrically related, and he suggested that 'it is the mother's metabolic turnover which, both in direct terms (through the physiology of gestation), and in indirect terms (through the partitioning of resources between maintenance and reproduction), determines the size of the neonate's brain' (Martin 1983, p. 14). However, it is entirely possible for two or more variables to be correlated with body weight and with each other, but to share no variation with each other independent of their relations with body weight. We analysed directly the relationship between neonatal brain weight and maternal metabolic turnover across mammals. Neonatal brain weight and maternal metabolism correlated highly, and were related with an allometric slope of 1.0 as Martin predicted (Pagel and Harvey 1988*b*). But, after controlling for the effects of adult body weight on both variables, their correlation was not significant.

We were also able to test directly Hofman's (1983*a*, *b*) prediction (based on a complicated set of substitutions among allometric relations) that maternal metabolic rate was the principal limiting factor of gestation length, but that variations in gestation length would lead to only small variations in neonatal brain weight. In fact, the data suggest just the opposite! Maternal metabolic rate was uncorrelated with species differences in gestation length, after controlling for adult body weight. But, even with adult body weight controlled, variation in gestation length was strongly correlated with variation in neonatal brain weight (Pagel and Harvey 1988*b*). This finding accords with the idea that mammal species with shorter gestation lengths simply complete less of their brain growth *in utero*, and have longer periods of brain growth after birth (Bennett and Harvey 1985*b*).

One attraction of combining allometric relations is the ability to produce novel predictions about data not yet in hand. However, even though combining allometric equations may be a useful exercise when thinking about the relations among a set of variables, the predictions should be regarded as highly tenuous until backed up by empirical evidence.

6.5 Techniques for fitting allometric equations

Three techniques are commonly used for estimating the parameters of the allometric equation $\log(Y) = \log(\alpha) + \beta \log(X)$: model I regression, major axis (principal axis or model II), and reduced major axis regression. These methods are appropriate for estimating β when influences on Y other than X are uncorrelated with X. We discuss in Section 6.8 the case of estimating β when there are correlated X variables. Comparisons of the methods from a variety of biological and statistical perspectives can be found in Kuhry and Marcus (1977), Sokal and Rohlf (1981), Harvey and Mace (1982), Seim and Saether (1983), and Pagel and Harvey (1988*a*).

Each of the three methods is derived from a more general model known as the general structural relations model (Sprent 1969; Rayner 1985) by making certain assumptions about the distribution of error in allometric data. The general structural relations model is the most general model for estimating the allometric line, and requires that the true and error variances of both variables are known, as well as the correlation of the errors. We assume for the rest of this discussion that the raw data have been logarithmically transformed before calculating variances and covariances. When Y and X are measured on the same specimens, correlations among error in Y and X are likely. However, for interspecific studies, where Y and X are likely to have been measured on different specimens, the assumption of uncorrelated error is probably reasonable. In such cases, the structural model (Sprent 1969; Kuhry and Marcus 1977; Rayner 1985) derived from the general structural model by assuming that the correlation of errors is zero, can be used. The structural model is the most general model for uncorrelated error, but requires that the ratio (denoted by λ) of the error variance in Y to that in X is known. Because error variances are seldom known, simplifying assumptions are usually made to derive estimators that can be applied to actual data (but see Section 6.5.1 below).

Model I regression is derived from the structural model by assuming that λ^{-1} is zero, which is equivalent to there being no error in the X variate. Model I is probably the most widely used technique for estimating the allometric line, and regression estimates have the useful property of minimizing the sum of the squared vertical deviations of the Y values about the regression line (Fig. 6.1). It is also the only technique that produces residuals (observed $Y-$ predicted Y) that are exactly uncorrelated with X. But for many problems model I is the least appropriate technique of the three because of its assumption of no error in X. The presence of error in X causes model I systematically to underestimate the true value of the slope.

Major axis and reduced major axis estimates allow error in both variates, a property better suited to most biological data. Major axis is derived from the structural relations model by assuming that $\lambda = 1.0$ (error in Y and X are equal), reduced major axis assumes that λ is equal to the ratio of the true variances in Y and X. Major axis analysis estimates the major axis of the bivariate ellipse formed by the joint distribution of Y and X, and produces a line that minimizes the sum of squared deviations perpendicular to itself (Fig. 6.1). A drawback of the major axis estimate of the slope is that it is not invariant to transformations of scale in Y and X. That is, the slope of the line does not change linearly with linear changes of scale. This is not a problem for data to be analysed in logarithmic form, however. Linear changes of scale in the raw data will be manifested as changes in the Y-axis intercept in the logarithmically transformed data.

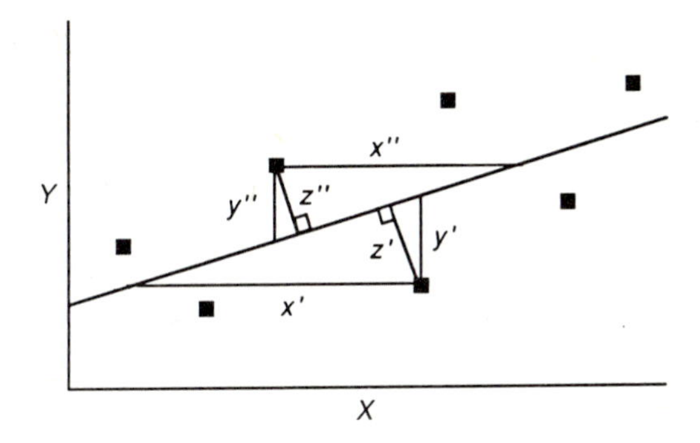

Fig. 6.1 Lines of best fit produced by a set of data (points are the black squares). Labelled lines drawn from two points show the distances minimized by regression analysis, major axis, and reduced major axis analysis (see text). Regression analysis of Y on X produces the line that minimizes the sum of the distances $(y')^2$ and $(y'')^2$; regression analysis of X on Y minimizes the sum of the distances $(x')^2$ and $(x'')^2$; major axis analysis minimizes the sum of the distances $(z')^2$ and $(z'')^2$; and reduced major axis minimizes the sum of the products $(x'y')$ and $(x''y'')$. Imbrie (1956) and Harvey and Mace (1982) give similar figures.

The reduced major axis estimator of the allometric slope is simply the ratio of the standard deviation of Y to the standard deviation of X (with a sign equal to the sign of the correlation between Y and X). Reduced major axis produces that line which minimizes the sum of the products of vertical multiplied by horizontal deviations of points from the line. However, it does not make use of any information about the covariance between Y and X in calculating the slope. Thus, it can yield nonsensical results, such as a slope between two variates that are uncorrelated, and for this reason we do not recommend its use.

The three techniques give different estimates of the slope unless the correlation coefficient is 1.0. Regression estimates are always shallower (for $r<1.0$) than major axis and reduced major axis, because of the (incorrect) assumption of no error in X. In practice, discrepancies among the models are small when the coefficient of determination (the square of the correlation coefficient) is greater than 0.90 to 0.95, but discrepancies increase as the coefficient of determination declines. Despite the undesirable property of underestimating the true slope, residuals from regression lines are, as we stated, uncorrelated with X. Residuals from major axis and reduced major axis are generally correlated with X, especially for lower values of the correlation. Thus, if the goal of an analysis is to control for the effects of weight, residuals calculated from a regression line should be used. However, if an accurate estimate of the true slope is the goal, major axis, reduced major axis, or the structural model is often preferable. Formulae for calculating slopes under the different statistical models together with worked examples are given in Box 6.1.

Box 6.1. Different methods for estimating the slope of a linear bivariate relationship

I. Formulae

Model I regression: $\hat{\beta} = \dfrac{\hat{\sigma}_{XY}}{\hat{\sigma}^2_X}$.

Major axis: $\hat{\beta} = \dfrac{\hat{\sigma}^2_Y - \hat{\sigma}^2_X + \sqrt{(\hat{\sigma}^2_Y - \hat{\sigma}^2_X)^2 + 4(\hat{\sigma}_{XY})^2}}{2\hat{\sigma}_{XY}}$.

Reduced major axis: $\hat{\beta} = \dfrac{\hat{\sigma}_Y}{\hat{\sigma}_X}$.

Structural relations: $\hat{\beta} = \dfrac{\hat{\sigma}^2_Y - \lambda\hat{\sigma}^2_X + \sqrt{(\hat{\sigma}^2_Y - \lambda\hat{\sigma}^2_X)^2 + 4\lambda(\hat{\sigma}_{XY})^2}}{2\hat{\sigma}_{XY}}$

where $\hat{\sigma}^2_X$, variance of X; $\hat{\sigma}^2_Y$, variance of Y;

$\hat{\sigma}_{XY}$, covariance of X and Y; $\lambda = (\hat{\sigma}^2_{eY})/(\hat{\sigma}^2_{eX})$,

where e = error.

II. Example

Assume: $\hat{\sigma}_X = 1.0$; $\hat{\sigma}_Y = 0.75$; $\hat{\sigma}_{XY} = r\hat{\sigma}_X\hat{\sigma}_Y$; $\sigma_{eYeX} = 0$.

	Correlation	
	0.70	0.90
Model I regression	0.525	0.675
Major axis	0.667	0.727
Reduced major axis	0.750	0.750
Structural relations with:		
$\lambda = 0.25$	0.879	0.782

$\lambda = (0.75)^2$	0.750	0.750
$\lambda = 1.00$	0.667	0.727
$\lambda = 1.5$	0.623	0.713
$\lambda^{-1} \rightarrow 0$	$\rightarrow$0.525	$\rightarrow$0.675

Note: The different methods make different assumptions about the value of λ. Model I regression assumes that $\lambda^{-1} = 0$, major axis assumes $\lambda = 1.0$, and reduced major axis assumes that $\lambda = (\hat{\sigma}^2_Y)/(\hat{\sigma}^2_X)$. Because we assume that errors are uncorrelated and that λ is known, the structural relations model gives the correct slope in each case. When the value of λ assumed by Model I regression, major axis or reduced major axis equals the true λ, then that method gives the same result as the structural relations model. For proof that the slope estimated by the structural relations model converges on that estimated by Model 1 regression as $\lambda^{-1} \rightarrow 0$, see Kuhry and Marcus (1977).

6.5.1 Estimating the error in the Y and X measures

The choice among one of the three commonly used allometric line fitting methods is usually a leap of faith rather than a rational judgement. Although it is easy to rule out Model I regression for estimating allometric slopes, how should we choose between major axis or reduced major axis? The answer is that we probably should not unless we have grounds for preferring one set of assumptions over the other. Instead of making a decision that amounts to a leap of faith, we can in most instances estimate the error directly by empirical means. Once we have such an estimate, the quantity λ can be substituted into the equation for the structural relations model to derive an estimate of the slope.

The quantities to be estimated are the error variances in Y and in X. Error variances, to a line fitting method, are all of those sources of variation that push points away from the allometric line. That is, the line fitting method assumes that if these sources of error could be removed, all of the points would fall along the line (this will not in general be true, a point we shall return to below). The line indicates the mean or expected value of, for example, Y at a given value of X, and the variation about that point, the error, is assumed to be normally distributed. If we knew where to draw the allometric line we could simply measure these sources of error directly. But that is the problem: we need to know the errors in order to know where to draw the line. Thus, we must come up with a way to estimate the errors independently of any particular line we might draw.

We have used two ways to estimate the variation of a species value (Pagel and Harvey 1988*c*, 1989*a*). The first method attempts to estimate the within population variance in Y and in X directly by measuring variation among individuals in a population. Then, the expected variation in the mean is just the variance of the population divided by the sample size. Thus, we estimated the standard error of the mean for brain and body weight in mammals as part of a simulation study of the tendency for allometric slopes to increase with taxonomic level. Data on individuals within a species were obtained for 23 different species from five orders and ranging in weight from mice (*Acomys*) to llama (*Lama*). The average variance in brain weight for these 23 species was about 1/5 of the average variance in body weight. This means that the expected sampling variation in species' mean brain weights can be expected to be about 1/5 as large as the sampling variation expected for mean body weights. Thus, λ was estimated as approximately 0.20.

Another way of estimating the variation in Y and X is sensitive to both within- and between-population differences. Instead of measuring the variation within populations, we estimated λ directly by studying the variation in the mean values reported in the literature for single species. Thus, we found 117 species for which brain and body weight values were available from two or more different sources, and calculated the variance in mean brain and body weight for each species. Then, λ was estimated as the mean of the variance in brain weight divided by the mean of the variance in body weight for the 117 species. This method gave an estimate for λ of 0.22, very similar to that found by the previous method.

The important point is that we were interested in developing estimates of the true allometric slopes. Critically, the value for λ that we found is very different from the value of λ that the three common line fitting methods assume. In this instance it meant that all three of these methods could be expected to underestimate the true allometric slope (Section 6.5.2).

The weakness of the approaches that we have described for estimating error is that there are reasons other than within- and between-population normal variation that account for species being found above or below allometric lines. For example, the small prosimian monkey *Daubentonia madagascariensis* (the aye-aye) has a very large brain for its body weight. Its large brain may be an adaptation to its particular lifestyle. This puts the aye-aye well above the allometric line relating brain weight to body weight for the other prosimians. Because sources of variation such as this one are not uniform across species, however, we have no way of measuring them directly. We can, however, remove this variation by statistical means or by removing certain groups from the data set so long as we have a way of identifying these groups objectively. For example, in our study of the scaling of mammalian brain weight on body weight, we controlled for the known associations between diet and relative brain weight both statistically and by removing some groups (Pagel and Harvey 1989*a*).

6.5.2 Effects of error on allometric slopes

The estimates derived from the three methods described above are only as good as the simplifying assumptions they make. A well-known phenomenon in the scaling of brain weight on body weight in mammals illustrates how evolutionary theory can sometimes be constructed on an edifice at least partly built from incorrect assumptions.

In studies of brain weight and body weight relations across species it has long been known that allometric slopes are generally higher when the species compared are more distantly related. Thus, among species in the same genus, slopes in the 0.20 to 0.40 range were often reported, whereas slopes linking species from different families or orders were often in the 0.67 to 0.75 range. Population-genetic and developmental-genetic models have been advanced to explain this pattern of slopes increasing with taxonomic level (Lande 1979; Riska and Atchley 1985). However, a pattern of slopes increasing with taxonomic level could arise as an artefact of using statistical models that make incorrect assumptions about the distribution of error in brain and body weight measurements. Our estimates discussed in Section (6.5.1) showed that the amounts of error in brain and body weight did not match the assumptions of regression, major axis or reduced major axis estimators. Further, the distribution of error was such that each of the techniques could be expected to underestimate the true slope, and the degree of underestimation would be greatest at lower taxonomic levels. This would produce a pattern of slopes increasing with taxonomic level. The reason why the methods are more accurate at higher taxonomic levels is that true variation increases among more distantly related taxa, while the amount of error stays the same or declines.

Based on our analysis of error patterns in the interspecific data, we conducted a simulation study of brain and body weight slopes. Simulated data for brain and body weight were generated for 1000 'species' assigned to arbitrary taxonomic groups. Error was added to the species points in quantities reflecting those in real data. Then slopes were estimated at successive taxonomic levels in the taxonomy. As predicted, each of the methods underestimated the true slope, and the degree of underestimation declined at higher taxonomic levels (Pagel and Harvey 1988*c*). We then used our estimates of error to analyse real data on brain and body weight in mammals (Pagel and Harvey 1989*a*). We employed the structural model described above to estimate the allometric slopes, using $\lambda = 0.20$ as estimated according to the procedures described above. Our analysis of over 900 species showed that there is no general tendency for slopes to increase with taxonomic level in the mammals. Those increases in slope with taxonomic level that were found tended to be associated with taxa that occupied particular ecological niches being more encephalized. It was already known that dietary differences were associated with differences in both body weight and encephalization in rodents, bats, and primates. And it was in precisely those orders that we found an increase in slope with the taxonomic level of analysis. When diet was controlled for, most of the taxonomic differences in encephalization vanished.

6.6 Applying the models for continuous variables to allometric relationships

The approach to estimating allometric slopes that we have described in the previous sections typically fits allometric lines to species. This was characteristic of most of the early work, and even much recent work, in allometry. Because of the growing recognition that species data points could not be considered independent for statistical purposes, following Clutton-Brock and Harvey (1977), allometric slopes have come increasingly often to be estimated from higher nodes, such as genera or families. But this approach, as discussed in Chapter 5, may also suffer from many of the same problems that plague across-species analyses. In this section we describe a new way of thinking about allometric slopes, based on the logic of independent comparisons that was developed in Chapter 5, that in principle gets around the problems of non-independence.

6.6.1 Fitting lines across species or higher nodes

There are two problems with fitting lines across taxa. One is that the species or higher nodes may not be independent, the other is that the allometric relationship *across* them may not be the same as that found independently *within* those hierarchically defined groups. The first problem we have dealt with in other guises throughout this book. The problems that non-independent data points pose for allometry are that standard errors will be underestimated as a result of overcounting the degrees of freedom, and the slope of the line may be distorted by particularly speciose taxa. Non-independence of data points means that we will place too much confidence in the estimates of slopes. This is not a trivial problem. For example, the debate over whether the correct slope of the line relating brain weight to body weight across taxa is 0.67 or 0.75 hinges partly on whose confidence intervals one believes. The issue of distortion means that we cannot trust the particular value of the slope as a description of all of the taxa.

The second major problem, that of across- versus within-taxa slopes, gets at the very heart of what we hope to estimate from the allometric relationship. Imagine two genera with two species each as shown in Fig. 6.2.

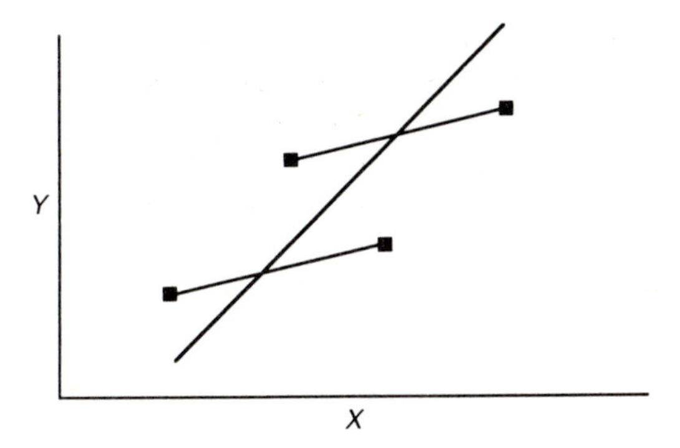

Fig. 6.2. The parallel lines each join species from the same genus. The steep line is a best fit line through the generic means on each axis.

The line relating Y to X is the same within each genus, but sharply steeper across all four species, or across the genera means. For purposes of the present discussion, we will call the slope estimated from the two within-genus relationships the *within-taxa slope* and the slope across the taxonomic groups the *across-taxa slope*. The within-taxa slope may tell us something about how two characters covary evolutionarily. The across-taxa slope will be sensitive to both the within-taxa slope and to vertical (in the two-coordinate diagram sense) displacements of one variable that are independent of changes in the other. The problem outlined in Fig. 6.2, the within-taxa and across-taxa slopes differing, may arise any time that allometric relationships are fitted across taxa which do not share the same immediate common ancestor. This is easily proved. The regression slope relating changes in Y to changes in X is given by the covariance between Y and X, divided by the variance of X. Thus,

$$\hat{\beta} = \frac{\hat{\sigma}_{xy}}{\hat{\sigma}_x^2} = \frac{\Sigma\,(X - \bar{X})\,(Y - \bar{Y})}{\Sigma\,(X - \bar{X})^2} \tag{6.3}$$

where $\hat{\beta}$ is the estimated regression coefficient. Consider that we are estimating $\hat{\beta}$ across the four species in Fig. 6.2. We can rewrite eqn (6.3) to express each X and Y, not simply as deviations about their respective overall means, but as follows:

$$\hat{\beta} = \frac{\Sigma\,[(X - \bar{X}_g) + (\bar{X}_g - \bar{X}_f)]\,[(Y - \bar{Y}_g) + (\bar{Y}_g - \bar{Y}_f)]}{\Sigma[\,(X - \bar{X}_g) + (\bar{X}_g - \bar{X}_f)\,]^2}. \tag{6.4}$$

Equation (6.4) simply expresses the deviation of X about the overall mean as a function of the deviation of each X about its genus mean ($\bar{X}_g$) plus the deviation of the genus mean about the family mean ($\bar{X}_{f\cdot}$). The same is done for Y. Of the four terms resulting from the product in the numerator, two, the cross products of genera with family values have expectations of zero. The cross product in the denominator also has expectation zero. Thus, eqn (6.4) simplifies to

$$\hat{\beta} = \frac{\Sigma(X - \bar{X}_g)(Y - \bar{Y}_g) + \Sigma(\bar{X}_g - \bar{X}_f)(\bar{Y}_g - \bar{Y}_f)}{\Sigma(X - \bar{X}_g)^2 + \Sigma(\bar{X}_g - \bar{X}_f)^2}. \tag{6.5}$$

Dividing top and bottom by the sample size (n - 1), the first two terms in the numerator are the covariance between X and Y within the genera and the covariance between the X and Y genera means within the family. The denominator becomes the variance of X within genera plus the variance of the X means within the family. Thus,

$$\hat{\beta} = \frac{\hat{\sigma}_{xy(\text{gen})} + \hat{\sigma}_{xy(\text{fam})}}{\hat{\sigma}^2_{x(\text{gen})} + \hat{\sigma}^2_{x(\text{fam})}}. \tag{6.6}$$

Equation (6.6) makes explicit the fact that the allometric slope estimated across species or higher nodes from more than one taxonomic group contains components of variation due to both within- and between-taxa effects. The components associated with families correspond to what we have termed the across-taxa slope, whereas the components associated with genera correspond to the within-taxa slope. Thus, the slope across taxa is a combination of the regression coefficient within taxa and the regression coefficient across taxa. From eqn (6.6) it is seen that only if the family components are proportional to the genus components will $\hat{\beta}$, the regression coefficient across taxa, be an estimate of the regression coefficient within taxa. This will not be true of the data in Fig. 6.2, but would be true of the data displayed in Fig. 6.3.

The conclusion to be drawn from eqns (6.3) to (6.6) is that whenever we attempt to estimate and interpret allometric slopes, we should take an explicit within-taxa versus across-taxa view. That is, we should not interpret the across-taxa slope as necessarily saying anything about the way that the two characters have shown correlated evolution within taxa. In the next section we describe a way of estimating allometric slopes from separate within-taxa effects.

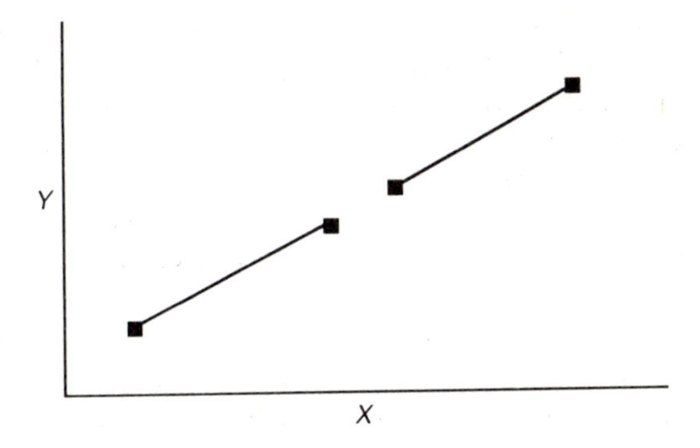

Fig. 6.3. The two lines linking species within genera have the same slope and intercept.

6.6.2 Allometry based on independent comparisons

The simplest way to take into account within-taxa versus across-taxa effects when doing allometry would be to do a nested analysis of covariance. This procedure would estimate the slope of the line separately within each of the lowest taxonomic groups (e.g. species within genera), then do the same at each higher level until, say, it estimated the slope among orders within the class. This is essentially the method that Bell (1989) advocates.

On the surface this technique solves the difficulties raised in the previous section: slopes are estimated separately at different taxonomic levels allowing explicit comparison and tests among taxonomic levels. However, it is the latter point, that of tests, that needs careful attention. In order to compare statistically the slopes from different taxonomic levels, or even slopes at the same level, we need an estimate of the standard error of the slope. And there's the rub because to calculate statistics on the slopes found within taxa relies on the assumption that the data points are independent. But following the logic of Chapter 5, we cannot routinely assume that the species within a genus, or more generally, the subtaxa within a higher taxon are independent. For example, imagine a genus containing 10 species, five with some character and five without it. It is probably not true that the character has evolved five times independently within the taxon. The same logic applies to continuous variables. The result is that, save for bifurcating phylogenies, the number of subtaxa within a taxon is very likely an overestimate of the true number of degrees of freedom. This means that the standard errors cannot be trusted.

One way around this problem is to treat the individual regression coefficients themselves as random variables and test whether the set of regression coefficients derived from, say, all within-genus comparisons differs from some null expectation, or from the value of the regression coefficients at some other taxonomic level. This test does not make use of the suspect standard errors, only the regression coefficients themselves. This is what Read (1987) and Pagel and Harvey (1989*a*) did. Although there is nothing wrong with this procedure it does have the limitation that there is no easy way to combine the data into an overall test of the estimate of the slope.

The procedure that we will describe below is designed to get around these limitations of nested-analysis-of-covariance-like analyses. In advocating independent-comparisons tests in Chapter 5 we argued that a proper test of a comparative relation relies on the number of times that the relation has evolved rather than on the number of species or other taxa that eventually inherit it. The logic of independent comparisons approaches is that within each taxon containing two or more subtaxa there is information bearing on the comparative relation. Here we seek a way of using that information to estimate and test allometric relations.

The independent-comparisons procedures described in Chapter 5 calculate a linear contrast or a weighted difference score on the X and Y variables within each taxonomic group. To review, the weighted difference score reduces the information on the n subtaxa within a taxon to a single data point. This single point represents our belief that the comparative relation has evolved at least once in that taxon. The crux of the method for estimating allometric lines that we shall develop here is that the set of linear contrast scores from the various taxa can be used to estimate the allometric slope. In fact, we shall prove below that given some assumptions about the way that comparisons are derived, the expected slope of the line based on the linear contrasts is equal to the true slope as estimated by traditional across-taxa approaches. The advantage of using comparisons, however, is that the data points are expected to be independent.

Consider an allometric relationship between Y and X such that $Y = \alpha + \beta X + e$, where α is the y-axis intercept, β is the slope of the line, and e represents the residual error. The slope β is usually estimated by regressing Y on to X across taxa. Here we want to prove that the slope β will also be estimated by regressing linear contrast scores in Y on to linear contrasts in X. A linear contrast in X is given by:

$$\Sigma c_i X_i = c_1 X_1 + c_2 X_2 + \ldots + c_n X_n. \qquad (6.7)$$

where the c_i are the contrast coefficients with the property that $\Sigma c_i = 0$, and the X_i are, in this instance, the species or other subtaxa values of X. Given the property that the contrast coefficients must sum to zero, the linear contrast defined in eqn (6.7) is a weighted difference score among the subtaxa. Now, the same linear contrast coefficients that were applied to the X_i are applied to the Y_i to get the corresponding contrast on Y:

$$\Sigma c_i Y_i = c_1 Y_1 + c_2 Y_2 + \ldots + c_n Y_n. \tag{6.8}$$

The linear relationship between Y and X means that we can substitute for Y in eqn (6.8) to get:

$$\Sigma c_i Y_i = \Sigma c_i(\alpha + \beta X_i + e_i) = \alpha \Sigma c_i + \beta \Sigma c_i X_i + \Sigma c_i e_i. \tag{6.9}$$

The first term on the right side of the equation is zero since α is a constant. Taking expectations, the last term on the right also goes to zero if the contrast coefficients are chosen independently of the residual variation in the Y variable. This is necessary to ensure that the expected covariance between the contrast coefficients and the residual errors is zero. The procedures by which the methods developed by Felsenstein (1985*a*), Grafen (1989), and Pagel and Harvey (1989*a*) choose values of the c_i were discussed in Chapter 5. If the first and third terms go to zero, then the expected value of eqn (6.9) is just $\beta \Sigma c_i X_i$, or β times the contrast on the X variables. The correlation between a variable and β times that variable will yield a line with a slope equal to β. Thus, the allometric slope β estimated from correlating a set of weighted difference scores will be an unbiased estimator of the true slope relating Y to X across taxa given the assumptions about the c_i. This means that we can use the linear contrasts from independent comparison procedures to estimate allometric slopes. As with any across-taxa slope, the within-taxa slope may contain both functional and phylogenetic components, but the advantages of this procedure are that the data points are independent, and the set of points can be used to test for an overall allometric relationship.

Two examples of allometry done on linear contrasts

The first example of estimating the allometric slope from linear contrasts comes from a data set on basal metabolic rate and body weight in mammals. Kleiber's (1961) well-known 'law' states that basal metabolic rate increases according to body weight raised to the 0.75 power, or, when both variables are plotted on logarithmic axes, the relationship should be a straight line with a slope of 0.75. Harvey *et al.* (1990) used Pagel and Harvey's independent-contrast method for continuous variables (see Chapter 5) to study the relationship of metabolic rate to body weight, as part of a larger study of the relationship of basal metabolic rate, body weight, and life history variation in mammals. Pagel and Harvey's (1989*a*) method chooses the c_i coefficients independently of the Y variable.

Independent contrasts were found at each level of a taxonomy including data from 18 different orders of mammals and 315 species.

Figure 6.4 plots the values of the independent comparisons on basal metabolic rate against the independent comparisons on body weight for the 90 independent comparisons available in that data set. Each pair of points on the graph plots what are essentially weighted change scores (change within a taxon). The set of changes in Y and changes in X should have a slope of approximately 0.75, according to Kleiber's law. In fact, the estimate of the slope is 0.75 (95% C.I. = 0.69–0.82) with a correlation of 0.92. (Using the structural relations model with $\lambda = 1.0$, equivalent to major axis, to estimate the slope).

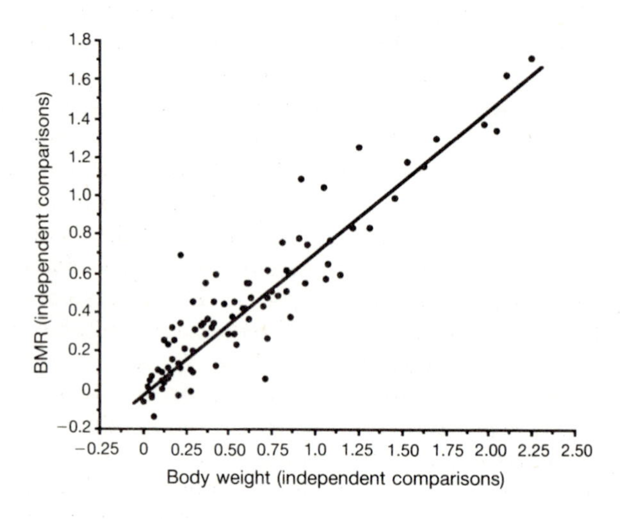

Fig. 6.4. Plot of independent contrasts on basal metabolic rate (BMR) versus independent contrasts on body weight (n = 90) obtained from an analysis of 315 mammal species. Each pair of points represents an independent comparison as described in the text. The slope of the line, fitted by the structural relations model with $\lambda = 1.0$, is approximately 0.75 (see text).

Just as with any other across-taxa information, it is possible that, along the lines of the argument developed in eqns (6.3) to (6.6), the slope of the line differs for different taxonomic levels. This is easily tested with the data in Fig. 6.4 because the independent comparisons can easily be grouped according to their taxonomic level. Thus, we can ask whether the slope across the independent comparisons derived from families differs from the independent comparisons derived from genera or orders. This is most easily done by using discrete codes in a multiple regression to identify the taxonomic level of a comparison. Then, after regressing the metabolic rate comparisons onto the body weight comparisons, one can ask whether information on taxonomic level significantly increases the amount of variance accounted for. In this case it does not (Table 6.1).

Table 6.1 Hierarchical regression of basal metabolic rate independent comparisons on to body mass independent comparisons, plus discrete variables coding for the taxonomic level of the comparison. The interaction of body weight with taxonomic level is coded for by the cross products of body weight and genus, and body weight and family. The absence of any effects due to the taxonomic level of the comparison means that the relationship between metabolic rate and body weight does not differ at different taxonomic levels. Note that the value of the regression coefficient changes slightly after controlling for taxonomic level.

Variable	Regression coefficient	*F* statistic
Body weight	0.78	119.8
Genus	0.03	0.1
Family	0.12	1.7
Body weight x genus	−0.07	0.4
Body weight x family	−0.13	2.4

Our second example comes from an analysis of brain weight and body weight in mammals. We analysed data from the 917 mammal species used by Pagel and Harvey (1989*a*), grouped according to eight taxonomic levels. We were able to calculate 308 separate independent comparisons on the data. From a number of analyses across-species or across-higher-taxa the slope of the line relating brain weight to body weight in mammals has been placed at between 0.67 and 0.75, with more recent studies favouring the higher figure. Recently, however, in an analysis of the tendency for slopes to increase with taxonomic level, we argued that assigning a single value to the brain weight–body weight relationship in mammals ignored the fact that there is substantial variation in the slope among orders: slopes across superfamilies or suborders within orders ranged from 0.49 in Artiodactyls to 0.83 in Primates (Pagel and Harvey 1989*a*). However, we concluded, there was no evidence for a general increase in slopes with taxonomic level *within* orders, once the effects of diet and ecology were controlled for. Thus, we argued, there is no reason to believe that the slope across orders of mammals (at about 0.75) described an evolutionary point towards which the various mammal orders were moving. Does this new type of analysis support our conclusions?

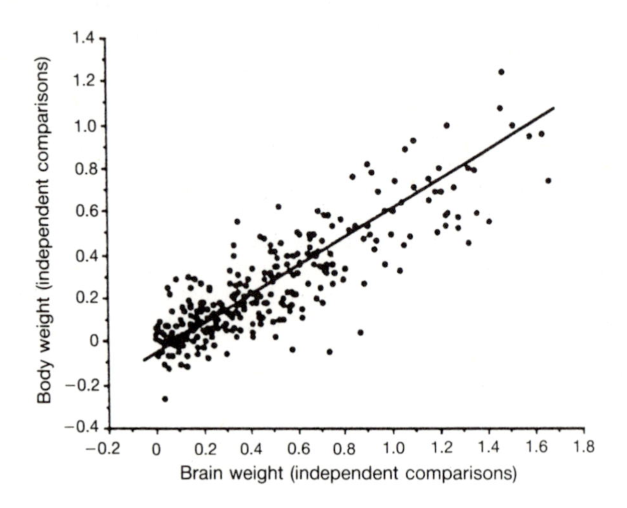

Fig. 6.5. Plot of independent contrasts on brain mass versus independent contrasts on body mass (n = 308) obtained from an analysis of 917 mammals. Each pair of points represents an independent comparison as discussed in the text. The slope of the line, fitted by the structural relations model with λ = 0.20, is approximately 0.69 (see text).

The 308 within-taxa contrasts on brain weight are plotted against the within-taxa body weight contrasts in Fig. 6.5. The slope of the line is 0.69 (95% C.I. = 0.65–0.73) by the structural relations model (λ = 0.20, see above). This indicates that, on average, unit increases in body weight within taxa are associated with 0.69 of a unit increase in brain weight. This figure is somewhat lower than the 0.75 figure often quoted for across-taxa results. The reason for this may be that the across-taxa result incorporates phylogenetic effects on the slope (see eqns 6.3 to 6.7 above) to a much greater extent than the present result. Unlike the metabolic rate example, there is substantial variation in the slopes among orders and among taxonomic levels, and thus we should not interpret the overall slope as a general description of mammals. Our previous conclusion is supported. Tables 6.2 and 6.3 present these data which, although analysed in a different way, are similar to those reported by Pagel and Harvey (1989*a*) who discuss the causes of the variation in slopes.

Table 6.2 Analysis of the relationship between brain and body mass in 917 mammals, using independent comparisons. Allometric slope is the slope of the brain weight independent comparisons on the body weight independent comparisons. All slopes were derived from the structural relations model (see text), with $\lambda = 0.20$. Sample size is the number of independent comparisons on which the slope is based. There is substantial heterogeneity among orders of mammals in the slopes linking brain mass comparisons to body mass comparisons.

Order	Allometric slope	Correlation coefficient	Sample size
Artiodactyla	0.54	0.87	31
Carnivora	0.70	0.90	42
Cetacea	0.46	0.83	18
Chiroptera	0.75	0.92	50
Insectivora	0.64	0.84	20
Lagomorpha	0.64	0.98	5
Marsupialia	0.70	0.96	18
Primates	0.92	0.91	42
Rodentia	0.56	0.89	67

Table 6.3 Analysis of the relationship between brain and body mass in 917 mammals, using independent comparisons. Allometric slope is the slope of the brain weight independent comparisons on the body weight independent comparisons. All slopes were derived from the structural relations model (see text), with $\lambda = 0.20$. Sample size is the number of independent comparisons on which the slope is based. There is heterogeneity among taxonomic levels in the slopes linking brain mass comparisons to body mass comparisons.

Taxonomic level	Allometric slope	Correlation coefficient	Sample size
Genera	0.66	0.82	154
Subfamilies	0.66	0.83	58
Families	0.67	0.84	53
Superfamilies	0.91	0.91	12
Suborders	1.00	0.78	11
Orders	0.68	0.93	11

6.7 Explaining allometry

The interpretation of allometric slopes, with a few notable exceptions (e.g. Alexander 1968, 1982; McMahon 1983), seldom goes beyond merely describing the slope of the line of some character with body weight, or of comparing two slopes and declaring that one of them shows more rapid change with evolutionary changes in body weight. This leaves allometry as little more than a description, although a useful one. Sometimes an investigator successfully predicts the value of an allometric coefficient from a collection of allometric slope coefficients that describe the web of interrelationships of several other variables. This gives us reason to believe that the correlations among those variables are all high, and may even suggest certain mechanisms. Nevertheless, these practices often do not give us insight into why the allometric coefficients take the values that they do in the first place. In this section we argue that allometry can begin to move away from this descriptive approach and towards attempts to explain allometric slopes in terms of the functions they are meant to represent, or in terms of the structure of environments. We give an example from life history variation in mammals.

Most of the variables related to the timing of life in mammals scale against body weight with an allometric slope of about 0.25, generally lying between 0.2 and 0.4 (Harvey and Read 1988). This curious finding extends right across the spectrum of life history characters including such characters as length of gestation, age at maturity, and inter-birth interval (Calder 1984; Harvey and Read 1988). Many physiological variables show a similar scaling (Calder 1984). Lindstedt (1985) and Lindstedt and Swain (1988) interpret this to imply the existence of an internal clock, which they christen the *Periodengeber*, that regulates all body processes around some internal metric set by body weight.

Lindstedt and Swain may turn out to be correct in their speculation about the existence of a *Periodengeber*. However, the difficulty with this sort of explanation is that on closer inspection it turns out not to be an explanation at all, but rather just a renaming of an empirical phenomenon. In contrast, there exists a well-developed body of theory on life history evolution (Charlesworth 1980; Partridge and Harvey 1988) that attributes variation in the timings of life to variation in the age-specific mortality schedules characteristic of a species. From this point of view, we ought to be able to predict a species' life history tactics from knowledge of the probabilities of survival to various ages, and the costs of reproduction. No internal clock is necessary for the explanation.

But why is it, then, that most species with short gestation lengths, early ages of maturity, and short inter-birth intervals are small? Demographic reality would predict that smaller animals must be those that suffer higher rates of age-dependent mortality. Birth rates must equal death rates over ecological time or species would be forever going extinct or their populations would be increasing indefinitely. This raises the possibility that something like the *Periodengeber* phenomenon might arise merely as a consequence of two other relationships. That is, the relationship of life history timings to body weight may simply follow from the fact that small animals suffer higher mortality, and higher rates of mortality must be compensated by increasing at least some components of fecundity (Sutherland *et al.* 1986; Harvey *et al.* 1989).

We can illustrate this idea with a data set compiled by Millar and Zammuto (1983) and subsequently used by Harvey and Zammuto (1985) and Sutherland *et al.* (1986). Millar and Zammuto collected together information on body weight, age at maturity, and life expectation from age at maturity for 29 mammal species. The subsequent papers showed that, using life expectation from birth or maturity as indicators of mortality rate, even after controlling for body weight, mammal species with shorter life expectations had earlier ages of maturity (see Chapter 5). These results fit nicely with our argument. However, we can also use the data to ask whether the relationships between age at maturity (M) and lifespan from maturity (L), and between lifespan from maturity and body weight (W) are such that we predict an exponent close to the empirical slope describing age at maturity and body weight.

We re-analysed the Millar and Zammuto (1983) data set using independent comparisons. This analysis yielded 23 independent comparisons from which we calculated the following relationships:

$$M \approx W^{0.30} \quad (6.10)$$

$$M \approx L^{1.17} \quad (6.11)$$

$$L \approx W^{0.21}. \quad (6.12)$$

Substituting eqn (6.12) into eqn (6.11), yields the following prediction:

$$M \approx (W^{0.21})^{1.17} = W^{0.25}. \quad (6.13)$$

This predicted result at 0.25 is close to the actual relationship of 0.30 between the two variables in this data set.

Our re-analysis of the Millar and Zammuto data set does not prove incorrect the *Periodengeber* hypothesis. Proponents of the *Periodengeber* would argue that it is in fact the internal clock that is setting these other relationships. But the point is that our way of looking at the scaling of life history variables does not rely on any hidden processes, and is derived from first principles.

We suggest, then, that the scaling of life history variables with body weight may derive from more fundamental relationships of weight with mortality schedules. But why should there be any consistent relationship between weight and rates of mortality? We do not yet know the answer to this question. It may be simply that some components of fecundity must compensate for differences in mortality in populations with density dependent regulation. We could then ask which components of fecundity might be expected to compensate, and examine whether species with high rates of natural mortality have evolved genetically determined differences in fecundity or whether fecundity compensates in response to local population density (Sæther 1988; Bennett and Harvey 1988; Promislow and Harvey 1990).

The critical point is that the demographic approach offers a way of explaining the value of allometric coefficients. For example, we do know that the relationship between body weight and the number of different species forms an approximate log-normal curve (Van Valen 1973; May 1978). That is, the number of different species declines sharply with increasing weight. This means that larger-bodied species will have fewer species of predators than will smaller bodied species. Larger-bodied species suffer lower rates of mortality from many other sources as well (see Clutton-Brock and Harvey 1983). For example, larger species are less susceptible to extremes of temperature, and can go longer without food.

There may be some very general principles, then, underlying the structure of environments that make body weight scale against mortality the ways that it does. This approach, of attempting to explain allometric relationships as a function of demographic reality, is, we think, preferable to the largely descriptive role that allometry has played over the years.

6.8 Allometric models with more than one independent variable

All of the statistical approaches described in this chapter have treated the relationship between Y and X as a simple bivariate one in which X is the only presumed influence on Y, or, if there are other influences on Y, they are presumed to be uncorrelated with X. Under these circumstances, the procedures described in the previous sections can be used to provide an unbiased estimate of the relationship between Y and X, given certain assumptions about the error variances in Y and X. In this section we briefly explore some of the consequences for estimating allometric relationships of situations in which the several presumed influences on Y cannot be assumed to be independent of each other. Thorough treatments of the issues explored below can be found in Blalock (1971) and Heise (1975). Riska (1991) provides an overview in the context of brain weight and body weight evolution.

Assume that we have two observed variables Y and X, both of which can be described as the sum of a true component and an error component:

$$Y = y_t + e_y. \quad (6.14)$$
$$X = x_t + e_x. \quad (6.15)$$

We leave the error terms unspecified except that we assume that they are uncorrelated with the true values, and that the two error terms are uncorrelated with each other. The regression coefficient of the true values, y and x, is given by the familiar ratio of the covariance of y and x to the variance of x:

$$\beta = \frac{\sigma_{x_t y_t}}{\sigma^2_{x_t}} \quad (6.16)$$

This value is estimated from the data by the ratio of the covariance of the observed Y and X scores to the variance of X:

$$\hat{\beta} = \frac{\sigma_{XY}}{\sigma^2_X} \quad (6.17)$$

The difficulty with eqn (6.17) arises from the fact that the expected value of $\hat{\beta}$ underestimates the true β:

$$E[\hat{\beta}] = E\left[\frac{\hat{\sigma}_{XY}}{\hat{\sigma}^2_X}\right] = \frac{\sigma_{x_t y_t}}{\sigma^2_{x_t} + \sigma^2_{e_x}} = \frac{\beta\, \sigma^2_{x_t}}{\sigma^2_{x_t} + \sigma^2_{e_x}} \quad (6.18)$$

Error in the X variable causes the estimator of the slope to be biased towards underestimating the true slope. The 'error' variance here is defined broadly to include any influence on X that is uncorrelated with x_t and y_t. There are various ways of correcting for this bias. One is to use statistical models that are designed to take into account error in both axes (i.e. in Y and X). These were discussed in the previous sections and are especially relevant when one is interested in describing the relationship between Y and X without designating one of the variables as the causal variable.

For instances in which X is treated as the presumed causal influence, one can attempt to estimate the magnitude of the error variance in X and subtract it from the denominator in eqn (6.18). This will give an unbiased estimate of the true slope when the causal model presumed to underlie variation in Y is that shown in Fig. 6.6.

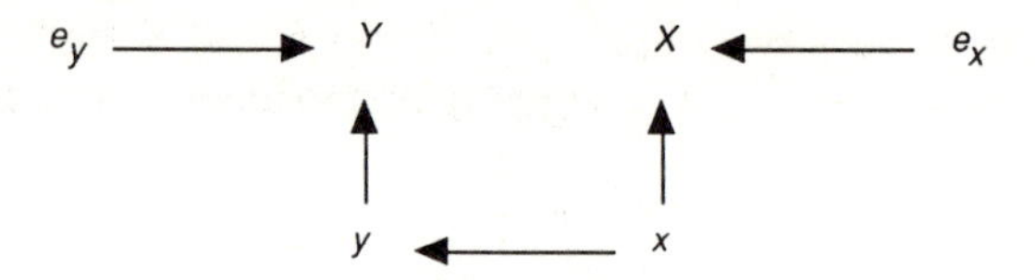

Fig. 6.6. Two observed variables Y and X represented as combinations of true and error components. The true values of y and x are possibly related, but the errors are assumed to be uncorrelated with each other and with their respective true values.

Consider, however, that variation in Y can be attributed to two sources X_1 and X_2, which may be correlated with each other, and both of which are measured with error. This scenario is depicted in Fig. 6.7.

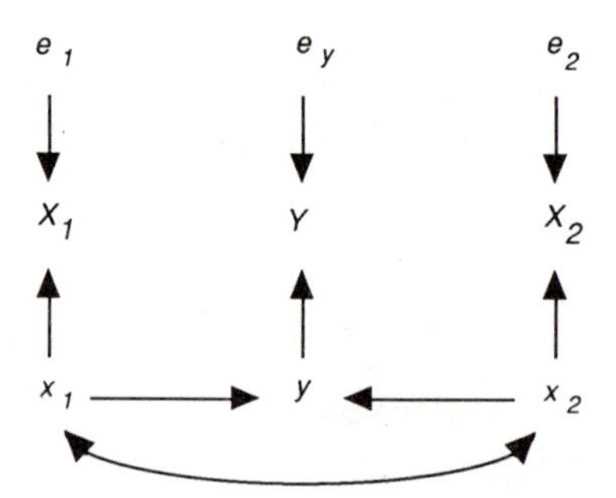

Fig. 6.7 An observed variable Y is portrayed as a function of true and error components. Variation in the true component of Y is the result of two variables, X_1 and X_2. In turn, X_1 and X_2 may be correlated.

We assume again that errors of measurement are uncorrelated with the true scores, and are uncorrelated with each other. Now if we estimate the slope of y on x_1 using eqn (6.17), we get the following. First replace Y with its description from the causal model of Fig. 6.7:

$$Y = \beta_1 x_1 + \beta_2 x_2 + e_y \, . \tag{6.19}$$

Each observed Y is presumed to be a function of the true x scores plus some error. Then, using COV and VAR to represent covariance and variance, respectively, the slope of the line relating the observed Y to X_1, as given by eqn (6.17) is:

$$\hat{\beta}_1 = \frac{\text{COV}\,(\beta_1 x_1 + \beta_2 x_2 + e_y\ X_1)}{\text{VAR}\,(X_1)} \tag{6.20}$$

The expected value of β_1 in eqn (6.20) is:

$$\text{E}[\hat{\beta}_1] = \frac{\beta_1 \sigma_{x_1}^2 + \beta_2 \sigma_{x_1 x_2}}{\sigma_{x_1}^2 + \sigma_{e_1}^2} \tag{6.21}$$

Equation (6.21) shows that the usual estimator of the slope relating Y and X_1 will be biased not just due to error in X_1, but also by any correlation between X_1 and X_2. On the other hand, if X_1 and X_2 are uncorrelated, eqn (6.21) reduces to eqn (6.18).

The result from eqn (6.21) shows that if Y is caused by more than one X variable, and the various causes of Y are correlated, the procedures for estimating the slopes are biased. For example, if X_1 and X_2 are positively related, if β_2 is positive, and if the error in X_1 is small, then the estimator in eqn (6.21) will be biased toward overestimating the true slope. This is very likely the reason why the allometric slopes describing the relationship of brain size to body size in mammals described in Sections 6.5.2 and 6.6.2., increase when the slopes are fitted across taxa from increasing higher taxonomic levels (Table 6.3). Distantly related taxa are very likely to differ in many more features than are closely related taxa. In the case of mammals, there appear to be dietary differences that are correlated with body weight which lead to higher slopes among families and superfamilies.

We have chosen a very simple presumed causal system for Fig. 6.7. Many others, including correlated error terms, are possible and quite likely to be true. Each different scenario leads to a slightly different set of biases. This kind of problem is widely discussed in the econometrics and sociological models literature, but has received very little attention in biology. Riska (1991), however, discusses one scenario of particular relevance to estimating the relationship between brain size and body size.

The key issue when confronted with a situation such as that in Fig. 6.7 is how to remove the effects of the correlation between X_1 and X_2 from the estimate of β_1. The trick is somehow to remove the covariance of x_1 and x_2 from the numerator of eqn (6.21), as well as to estimate the error in X_1. The trouble with using standard partial correlation techniques is that they do not 'know' how to partition the shared variance between X_1 and X_2. However, it is possible to get around this problem by finding a variable or set of variables that predict X_1, but which are uncorrelated with X_2. These variables are called 'instrumental variables' and when used to predict X_1, create a new X_1 which, in theory, is uncorrelated with X_2. Then, the relationship of Y to the new X_1 can be studied without bias creeping in due to the relationship between X_1 and X_2. Blalock (1971) and Heise (1975) provide general introductions to the use of instrumental variables, and Riska (1991) discusses these approaches with reference to the brain size and body size issue. A difficulty with the instrumental variables approach is that it may often be difficult to find good instrumental variables. It is essential that they are uncorrelated with the other measured variables, and to be efficient they must also predict a substantial proportion of the variance in the variable for which they are used as instrumental variables. Nevertheless, the issues raised by the problem of correlated causes need attention.

An alternative to the instrumental variables approach is to select units for study in such a way as to control for possible correlated influences on Y. For example, among the bats, the Microchiroptera are small and tend to eat insects or feed on nectar, whereas the Megachiroptera are big and feed on fruit. An allometric slope relating brain size to body size fitted to data from both groups confounds the influences of diet and body size and, in this case, results in a slope biased upwards. The slope can, however, be estimated separately within each group, thereby controlling for the influences of diet. This approach avoids the difficulties of finding instrumental variables for body size, but depends upon the validity of the assumption that diet and body size are the only other important influences on brain size.

6.9 Summary

The form of a comparative relationship can give insight into the reasons for its existence. In this chapter we use allometry as a case study to illustrate methods that can be used to determine the form of functional relationships. We show how the techniques developed in the previous chapter, for incorporating phylogenetic information into statistical tests for the existence of comparative relationships, can be extended to reveal the form of those relationships. The new methods are applied to particular data sets to examine the allometry of brain size, basal metabolic rates, mortality patterns, and life history variation. We show how such analyses can be used to move the study of allometry, and that of other functional relationships, from an exercise of description to one of interpretation.

7

Conclusion

The comparative method that we have described in this book depends on the interplay among phylogenetic inheritance, chance events, and adaptation to contemporary environments. To consider only one of these processes alone is to lose much that has intrigued evolutionary biologists since Darwin. Yet, although the comparative method has been in constant use over the past century or so, we can only echo Ridley's (1983*a*) wonder that it remained virtually unchanged from Darwin's time until a decade ago. Meanwhile, population genetics and experimental investigations had forged ahead. We hope that this book has shown why the quotation which opened our first chapter should soon become history: . . . 'we must learn to treat comparative data with the same respect as we would treat experimental results' (Maynard Smith and Holliday 1979, p. vii). We can already treat properly analysed comparative results with the same respect that we treat experimental results. And, as we have seen, each has a crucial role to play in the development of our field.

We have stressed throughout the important role that models of evolutionary change play in our statistical methods. Brownian motion models have been put to use for characterizing change in continuously varying characters, as has a Markov model in the case of dichotomous characters. New models, based on undoubtedly wicked mathematics, will gradually emerge. Two interconnected areas whose applications to comparative methods are moving ahead rapidly are phylogeny reconstruction and ancestral character state reconstruction. There is uncertainty in any reconstruction of a phylogeny or the ancestral states at its nodes. Recall that if transition probabilities among character states are greater than zero, any phylogeny or set of ancestral character states is possible, some are just more likely than others. Comparative methods need to be developed that can take into account the uncertainty about the phylogeny. Some comparative methods depend upon a single reconstruction of the character states at higher nodes. Here again, we need methods that acknowledge the uncertainty and can either gauge its effect on the results, or incorporate the uncertainty into the test. It may even be possible to develop methods that achieve both tasks by incorporating into a single test the uncertainty about the phylogeny and about the ancestral characters states.

There is still much to be done but we suspect that the groundwork for the future has been laid. The recognition and detection of independent evolutionary events has as much interest for paleontologists and taxonomists as it has for those of us who seek the reasons for organic diversity. These are exciting times, not least because the new wave of molecular genetic information has the potential to heal the schism that has increasingly separated molecular and organismic biologists in recent years. Although molecular information still provides us with character states, and the old problems of interpretation remain, the nature and precision of molecular data should provide us with ever more accurately constructed phylogenetic trees. It should provide us with new levels of organic diversity, some of which may be correctly interpreted as adaptive (Kreitman 1987). Perhaps we should start by asking 'How common *is* molecular convergence?'

Although molecules have much to give to and take from the comparative method, so do all points along the biological spectrum, even to the extreme of ecology. For example, the analysis of patterns in community structure may yet provide us with an increased understanding of the way phenotypes respond to ever changing environments, and which new niches are invaded and why (Sugihara 1980). If we are to retrace the reasons for evolutionary change, so vital to the proper functioning of the comparative method, those are the questions to which we must have answers. But, surprisingly perhaps, community analyses have yet to incorporate the comparative lesson that common phylogenetic inheritance dictates non-independence of species' values (Harvey and Pagel 1989). Just as species cannot be considered as statistically independent points for the comparative analyses described in this book, so a proper understanding of community structure will not be achieved independently of phylogenetic structure. Notable exceptions to the lack of work in this area are the 'null model' studies of community structure which have, on occasion, incorporated taxonomic affinity as a variable in both simulation models and statistical analyses (see Graves and Gotelli 1983; Harvey *et al.* 1983; Colwell and Winkler 1984). Ecological guilds contain species that may or may not be closely related. If guilds are the natural units of community function, their phylogenetic composition merits proper study.

What of evolution itself? The methods developed in this book seek as much that is unknown as they use what is known. All comparative analyses, as we have frequently stressed, are critically dependent on models of the ways that evolution proceeds. Until we know which models are the more appropriate, we shall not be sure of our conclusions. As the range of relevant models becomes more tightly circumscribed, so will our conclusions. Nowadays the limiting factor is not the range of available models, but knowing which one to choose. The choice comes down to deciding which model makes the most realistic assumptions about the evolutionary process. Tell us how to reconstruct the past, and we shall perform the comparative analysis with precision.

References

Adams, J. (1985). The definition and interpretation of guild structure in ecological communities. *Journal of Animal Ecology* **54,** 43–59.

Alberch, P. and Gale, E. (1983). Size dependence during the development of the amphibian foot. Colchicine-induced digital loss and reduction. *Journal of Embryology and Experimental Morphology* **76,** 177–97.

Alexander, R. D., Hoogland, J. L., Howard, R., Noonan, K. M., and Sherman, P. W. (1979). Sexual dimorphism and breeding systems in pinnipeds, ungulates, primates and humans. In *Evolutionary biology and human social behaviour* (ed. N. A. Chagnon and W. D. Irons), pp. 402–35, Duxbury Press, North Scituate, Mass.

Alexander, R. M. (1968). *Animal mechanics.* University of Washington Press, Seattle.

Alexander, R. M. (1982). *Optima for animals.* Arnold, London.

Aristotle (1982). *De partibus animalium.* Oxford University Press, Oxford.

Armstrong, E. (1983). Relative brain size and metabolism in mammals. *Science* **220,** 1302–4.

Avise, J. C., Neigel, J. E. and Arnold, J. (1984). Demographic influences on mitochondrial DNA lineage survivorship in animal populations. *Journal of Molecular Evolution* **20,** 99–105.

Avise, J. C., Arnold, J., Ball, J. M., Bermingham, E., Lamb, T., Neigel, J. E., Reeb, C. A. and Saunders, N. C. (1987). Intraspecific phylogeography: the mitochondrial bridge between population genetics and systematics. *Annual Review of Ecology and Systematics* **18,** 489–522.

Bell, G. (1982). *The masterpiece of nature: the evolution and genetics of sexuality.* Croom Helm, London.

Bell, G. (1989). A comparative method. *American Naturalist* **133,** 553–71.

Bennett, P. M. and Harvey, P. H. (1985a). Relative brain size and ecology in birds. *Journal of Zoology* **207,** 151–69.

Bennett, P. M. and Harvey, P. H. (1985b). Brain size, development and metabolism in birds and mammals. *Journal of Zoology* **207,** 491–509.

Bennett, P. M. and Harvey, P. H. (1987). Active and resting metabolism in birds: allometry, phylogeny and ecology. *Journal of Zoology (London)* **213,** 327–63.

Bennett, P. M. and Harvey, P. H. (1988). How fecundity balances mortality in birds. *Nature* **333,** 216.

Benton, M.J. (1988*a*) (ed.). *The phylogeny and classification of mammals.* Vol. 2, *Mammals.* Systematics Association Special Volume 35B. Oxford.

Benton, M.J. (1988*b*). The relationships of the major groups of mammals. *Trends in Ecology and Evolution* **3,** 40–45.

Berry, J. F. and Shine, R. (1980). Sexual size dimorphism and sexual selection in turtles. *Oecologia* **44,** 185–91.

Blackburn, T. M. (in press). The interspecific relationship between egg size and clutch size in waterfowl—a reply to Rohwer. *Auk.*

Blackburn, T.M., Harvey, P.H. and Pagel, M.D. (1990). Species number, population density and body size relationships in natural communities. *Journal of Animal Ecology* **59,** 335–45.

Blalock, H.M. ed. (1971). *Causal models in the social sciences.* MacMillan, London.

Bock, W. J. (1977). Adaptation and the comparative method. In *Major patterns in vertebrate evolution* (ed. M. K. Hecht, P. Goody, and B. Hecht), pp. 57–82, Plenum Press, New York.

Bock, W. J. (1980). The definition and recognition of biological adaptation. *American Zoologist* **20,** 217–27.

Bonner, J.T. (1965). *Size and cycle.* Princeton University Press, New Jersey.

Boyd, R. and Richerson, D. (1986). *Culture and the evolutionary process.* University of Chicago Press.

Bradley, J. V. (1968). *Distribution-free statistical tests.* Prentice Hall, Englewood Cliffs, New Jersey.

Brandon, R. N. (1978). Adaptation and evolutionary theory. *Studies in the History and the Philosophy of Science* **9,** 181–206.

Britten, R. J. (1986). Rates of DNA sequence evolution differ between taxonomic groups. *Science* **231,** 1393–98.

Brody, S. (1945). *Bioenergetics and growth.* Hafner, New York.

Brody, S., Procter, R. C., and Ashworth, U. S. (1934). Basal metabolism, endogenous nitrogen, creatinine and neutral sulphur excretions as functions of body weight. *University of Missouri Agricultural Experimental Station Research Bulletin* **89,** 1–18.

Brooks, D. R., O'Grady, R. T., and Glen, D. R. (1985*a*). The phylogeny of the Cercomeria Brooks, 1982 (Platyhelminthes). *Proceedings of the Helminthological Society of Washington* **52,** 1–20.

Brooks, D. R., O'Grady, R. T., and Glen, D. R. (1985*b*). Phylogenetic analysis of the Digenea (Platyhelminthes: Cercomeria) with comments on their adaptive radiation. *Canadian Journal of Zoology* **63,** 411–43.

Brown, W. M., Prager, E. M., Wang, A., and Wilson, A. C. (1982). Mitochondrial DNA sequences from primates: tempo and mode of evolution. *Journal of Molecular Evolution* **18,** 225–39.

Bull, J. J. and Charnov, E. L. (1985). Irreversible evolution. *Evolution* **39,** 1149–55.

Bull, J. J. and Charnov, E. L. (1988). How fundamental are Fisherian sex ratios? *Oxford Surveys in Evolutionary Biology* **5,** 96–135.

Burke, T. (1989). DNA fingerprinting and other methods for the study of mating success. *Trends in Ecology and Evolution* **4,** 139–44.

Burt, A. (1989). Comparative methods using phylogenetically independent contrasts. *Oxford Surveys in Evolutionary Biology* **6,** 33–53.

Burt, A. and Bell, G. (1987). Mammalian chiasma frequencies as a test of two theories of recombination. *Nature* **326,** 803–805.

Cain, A. J. (1964). The perfection of animals. *Viewpoints in Biology* **3,** 36–63.

Calder, W. A. (1984). *Size, function and life history*. Harvard University Press, Boston, Mass.

Camin, J. H. and Sokal, R. R. (1965). A method for deducing branching sequences in phylogenies. *Evolution* **19,** 311–26.

Carpenter, J. M. (1989). Testing scenarios: wasp social behavior. *Cladistics* **5,** 131–144.

Carson, H. (1983). Chromosomal sequences and inter-island colonizations in Hawaiian *Drosophila*. *Genetics* **103,** 465–82.

Carson, H. L. and Kaneshiro, K. Y. (1976). *Drosophila* of Hawaii: systematics and evolutionary genetics. *Annual Review of Ecology and Systematics* **7,** 311–46.

Carson, H. L., Clayton, F. E., and Stalker, H. D. (1967). Karyotypic stability and speciation in Hawaiian *Drosophila*. *Proceedings of the National Academy of Sciences (U.S.A.)* **57,** 1280–85.

Cavalli-Sforza, L. L. and Edwards, A. W. F. (1964). Analysis of human evolution. *Proceedings of the Eleventh International Congress on Genetics* **3,** 923–33.

Cavalli-Sforza, L. L. and Edwards, A. W. F. (1967). Phylogenetic analysis: models and estimation procedures. *American Journal of Human Genetics* **19,** 233–57.

Cavalli-Sforza, L. L., Barrai, I., and Edwards, A. W. F. (1964). Analysis of human evolution under random genetic drift. *Cold Spring Harbor Symposia on Quantitative Biology* **29,** 9–20.

Cavender, J. A. (1978). Taxonomy with confidence. *Mathematical Bioscience* **40,** 271–280; and (1979) *erratum, Mathematical Bioscience* **44,** 308.

Charlesworth, B. (1980). *Evolution in age-structured populations*. Cambridge University Press, Cambridge.

Charnov, E. L. (1982). *The theory of sex allocation*. Princeton University Press, New Jersey.

Cheverud, J. M. (1982). Phenotypic, genetic and environmental morphological integration in the cranium. *Evolution*, **36,** 499–516.

Cheverud, J. M., Dow, M. M., and Leutenegger, W. (1985). The quantitative assessment of phylogenetic constraints in comparative analyses: sexual dimorphism in body weight among primates. *Evolution* **39,** 1335–51.

Chew, F. S. and Robbins, R. K. (1984). Egg-laying in butterflies. In *The biology of butterflies* (ed. R. I. Vane-Wright and P. C. Ackery), pp. 65–79. Academic Press, London.

Clutton-Brock, T. H. (1986). Sex ratio variation in birds. *The Ibis* **128,** 329.

Clutton-Brock, T.H. (1989). Female transfer and inbreeding avoidance in social mammals. *Nature* **337,** 70–2.

Clutton-Brock, T. H., Albon, S. D., and Harvey, P. H. (1980). Antlers, body size and breeding group size in the Cervidae. *Nature* **285,** 565–7.

Clutton-Brock, T. H. and Harvey, P. H. (1977). Primate ecology and social organisation. *Journal of Zoology* **183,** 1–33.

Clutton-Brock, T. H. and Harvey, P. H. (1979). Comparison and adaptation. *Proceedings of the Royal Society of London, Series B* **205,** 547–65.

Clutton-Brock, T. H. and Harvey, P. H. (1983). The functional significance of variation in body size among mammals. In *Advances in the study of mammalian behavior* (ed. J. F. Eisenberg and D. G. Kleiman), pp. 632–63. American Society of Mammalogists, New York.

Clutton-Brock, T. H. and Harvey, P. H. (1984). Comparative approaches to investigating adaptation. In *Behavioural ecology: an evolutionary approach* (2nd edn), (ed. J. R. Krebs and N. B. Davies), pp. 7–29. Blackwell, Oxford.

Clutton-Brock, T. H. and Iason, G. R. (1986). Sex ratio variation in mammals. *Quarterly Review of Biology* **61,** 339–74.

Clutton-Brock, T. H., Harvey, P. H., and Rudder, B. (1977). Sexual dimorphism, socionomic sex ratio, and body weight in primates. *Nature* **269,** 797–800.

Coddington, J. A. (1986*a*). The monophyletic origin of the orb web. In *Spider webs and spider behavior* (ed. W. A. Shear), pp. 319–63, Stanford University Press, Palo Alto, California.

Coddington, J. A. (1986*b*). The genera of the spider family Theridiosomatidae. *Smithsonian Contributions to Zoology* **422,** 1–96.

Coddington, J. A. (1988). Cladistic tests of adaptational hypotheses. *Cladistics* **4,** 3–22.

Colwell, R. K. and Winkler, D. W. (1984). A null model for null models in biogeography. In *Ecological communities: conceptual issues and the evidence* (ed. D. R. Strong, D. Simberloff, L. G. Abele, and A. B. Thistle), pp. 344–59, Princeton University Press, New Jersey.

Cope, E. D. (1885). On the evolution of the vertebrata, progressive and retrogressive. *American Naturalist* **19,** 140–353.

Corbet, G.B. and Hill, J.E. (1980). *A world list of mammalian species.* Comstock Associates, Cornell University Press, Ithaca.

Crisci, J. V. and Stuessy, T. F. (1980). Determining primitive character states for phylogenetic reconstruction. *Systematic Botany* **5,** 112–35.

Crook, J. H. (1964). The evolution of social organization and visual communication in the weaver birds (Ploceinae). *Behaviour (Suppl.)* **10,** 1–178.

Crook, J. H. (1965). The adaptive significance of avian social organization. *Symposia of the Zoological Society of London* **14,** 181–218.

Crook, J. H. and Gartlan, J. S. (1966). Evolution of primate societies. *Nature* **210,** 1200–3.

Crow, J. F. and Kimura, M. (1970). *An introduction to population genetics theory.* Harper and Row, New York.

Damuth, J. (1981). Population density and body size in mammals. *Nature* **290,** 699–700.

Damuth, J. (1987). Interspecific allometry of population density in mammals and other animals: the independence of body mass and population energy-use. *Biological Journal of the Linnean Society* **31,** 193–246.

Darwin, C. (1859). *On the origin of species by means of natural selection, or, the preservation of favoured races in the struggle for life.* John Murray, London.

Darwin, C. (1871). *The descent of man and selection in relation to sex.* John Murray, London.

Davies, N. B. and Houston, A. I. (1984). Territorial economics. In *Behavioural ecology: an evolutionary approach* (ed. J. R. Krebs and N. B. Davies), pp. 148–69, Blackwell, Oxford.

DeVore, I. and Hall, K. R. L. (1965). Baboon ecology. In *Primate behavior* (ed. I. DeVore), pp. 20–52, Holt, Rinehart, and Winston, New York.

Diamond, J. M. (1983). Taxonomy by nucleotides. *Nature* **305,** 17–8.

Diamond, J.M. and May, R.M. (1977). Species turnover rates on islands: dependence on census interval. *Science* **197,** 266–70.

Donoghue, M. J. (1989). Phylogenies and the analysis of evolutionary sequences, with examples from seed plants. *Evolution* **43,** 1137–56.

Draper, N. and Smith, H. (1981). *Applied regression analysis.* Wiley, New York.

Dubois, E. (1898). Über die Abhängigkeit des Hirngewichtes von der Körpergrösse bei den Säugetieren. *Archiv für Anthropologie und Geologie Schleswig-Holsteins und der benachbarten Gebiete* **25,** 1–28.

Dumont, J. P. C. and Robertson, R. M. (1986). Neuronal circuits: an evolutionary perspective. *Science* **233,** 849–53.

Dunbar, R. I. M. (1982). Adaptation, fitness and the evolutionary tautology. In *Current problems in sociobiology* (ed. King's College Sociobiology Group), pp. 9–28, Cambridge University Press.

Dunham, A.E. and Miles, D.B. (1985). Patterns of covariation in life history traits of squamate reptiles: the effects of size and phylogeny reconsidered. *American Naturalist* **126,** 231–257.

Edmunds, M. (1974). *Defence in animals.* Longman, Essex.

Edwards, A. W. F. and Cavalli-Sforza, L. L. (1964). Reconstruction of evolutionary trees. In *Reconstruction of evolutionary trees* (ed. V. H. Heywood and J. McNeill), pp. 67–76, Systematics Association, London.

Efron, B. and Gong, G. (1983). A leisurely look at the bootstrap, the jacknife, and cross-validation. *American Statistician* **37,** 36–48.

Eisenberg, J. F. (1981). *The mammalian radiations.* The Athlone Press, London.

Eldred, A. K. (1991). Burt and Bell reanalyzed. *Nature.* (In press.)

Elgar, M. A. and Harvey, P. H. (1987). Basal metabolic rates in mammals: allometry, phylogeny and ecology. *Functional Ecology* **1,** 25–36.

Elgar, M. A., Pagel, M. D., and Harvey, P. H. (1988). Sleep in mammals. *Animal Behaviour* **36,** 1407–19.

Endler, J. A. (1986). *Natural selection in the wild.* Princeton University Press, New Jersey.

Estabrook, G. F. (1972). Cladistic methodology: a discussion of the theoretical basis for the induction of evolutionary history. *Annual Review of Ecology and Systematics* **3,** 427–56.

Estabrook, G. F. (1980). The compatibility of occurrence patterns of chemicals in plants. In *Chemosystematics: principles and practice* (ed. F. A. Bisby, J. C. Vaughan, and C. A. Wright), pp. 379–97, Academic Press, London.

Estabrook, G. F. and Anderson, W. R. (1978). An estimate of phylogenetic relationships within the genus *Crusea* (Rubiaceae) using character compatibility analysis. *Systematic Botany* **3,** 179–96.

Estabrook, G. F., Strauch Jr, J. G. and Fiala, K. L. (1977). An application of compatibility analysis to the Blackiths' data on orthopteroid insects. *Systematic Zoology* **26,** 269–76.

Esteal, S. (1990). The pattern of mammalian evolution and the relative rate of molecular evolution. *Genetics* **124,** 165–73.

Falconer, D. S. (1981). *An introduction to quantitative genetics.* Longman, London.

Farris, J. S. (1970). Methods for computing Wagner trees. *Systematic Zoology* **19,** 83–92.

Farris, J. S. (1978). Inferring phylogenetic trees from chromosome inversion data. *Systematic Zoology* **27,** 275–84.

Felsenstein, J. (1973*a*). Maximum-likelihood estimation of evolutionary trees from continuous characters. *American Journal of Human Genetics* **25,** 471–92.

Felsenstein, J. (1973*b*). Maximum-likelihood and minimum-steps methods for estimating evolutionary trees from data on discrete characters. *Systematic Zoology* **22,** 240–9.

Felsenstein, J. (1977). Excursions along the interface between disruptive and stabilising selection. *Genetics* **93,** 773–95.

Felsenstein, J. (1978). Cases in which parsimony or compatibility methods will be positively misleading. *Systematic Zoology* **27,** 401–10.

Felsenstein, J. (1979). Alternative methods of phylogenetic inference and their interrelationships. *Systematic Zoology* **28,** 49–62.

Felsenstein, J. (1981*a*). A likelihood approach to character weighting and what it tells us about parsimony and compatibility. *Biological Journal of the Linnaean Society* **16,** 183–196.

Felsenstein, J. (1981*b*). Evolutionary trees from DNA sequences: a maximum likelihood approach. *Journal of Molecular Evolution* **17,** 368–376.

Felsenstein, J. (1982). Numerical methods for inferring evolutionary trees. *Quarterly Review of Biology* **57,** 379–404.

Felsenstein, J. (1983). Parsimony in systematics: biological and statistical issues. *Annual Review of Ecology and Systematics* **14,** 313–33.

Felsenstein, J. (1984). Distance methods for inferring phylogenies: a justification. *Evolution* **38,** 16–24.

Felsenstein, J. (1985*a*). Phylogenies and the comparative method. *American Naturalist* **125,** 1–15.

Felsenstein, J. (1985*b*). Phylogenies from gene frequencies: a statistical problem. *Systematic Zoology* **43,** 300–11.

Felsenstein, J. (1985*c*). Confidence limits on phylogenies: an approach utilizing the bootstrap. *Evolution* **39,** 783–91.

Felsenstein, J. (1986). Waiting for post-neo-Darwinism. *Evolution* **40,** 883–9.

Felsenstein, J. (1988). Phylogenies and quantitative methods. *Annual Review of Ecology and Systematics* **19,** 445–71.

Fink, W. L. (1982). The conceptual relation between ontogeny and phylogeny. *Paleobiology* **8,** 254–64.

Fisher, R. A. (1930). *The genetical theory of natural selection*. Clarendon Press, Oxford.

Fitch, W. M. (1971). Toward defining the course of evolution: minimal change for a specific tree topology. *Systematic Zoology* **20,** 406–16.

Fitch, W. M. and Margoliash, E. (1970). Construction of phylogenetic trees. *Science* **155,** 279–84.

Fleagle, J. G. (1985). Size and adaptation in primates. In *Size and scaling in primate biology* (ed. W. L. Jungers), pp. 1–19, Plenum Press, New York.

Friday, A. (1989). Quantitative aspects of the estimation of evolutionary trees. *Folia primatologica* **53,** 221–234.

Futuyma, D. J. (1979). *Evolutionary biology*. Sinauer, Sunderland, Mass.

Futuyma, D. J. (1986). *Evolutionary biology*, (2nd edn). Sinauer, Sunderland, Mass.

Garland, T. and Huey, R. B. (1987). Testing symmorphosis: does structure match functional requirements? *Evolution* **4,** 1404–49.

Gerber, H. S. and Klostermeyer, E. C. (1970). Sex control by bees: a voluntary act of egg fertilization during oviposition. *Science* **167,** 82–4.

Ghiselin, M. T. (1972). Models in phylogeny. In *Models in paleobiology* (ed. T. J. M. Schopf), pp. 130–45. Freeman and Cooper, San Francisco.

Ghiselin, M. T. (1988). The origin of molluscs in the light of molecular evolution. *Oxford Surveys in Evolutionary Biology* **5,** 66–95.

Gillespie, J. H. (1986*a*). Rates of molecular evolution. *Annual Review of Ecology and Systematics* **17,** 637–65.

Gillespie, J. H. (1986*b*). Variability of evolutionary rates of DNA. *Genetics* **113,** 1077–91.

Gillespie, J. H. (1987). Molecular evolution and the neutral allele theory. *Oxford Surveys in Evolutionary Biology* **4,** 10–37.

Gittleman, J. L. (1981). The phylogeny of parental care in fishes. *Animal Behaviour* **29,** 936–41.

Gittleman, J. L. (1983). The behavioural ecology of carnivores. D. Phil. thesis, University of Sussex, Brighton.

Gittleman, J.L. (1985). Communal care in mammals. In *Evolution: essays in honour of John Maynard Smith* (ed. P.J. Greenwood, P.H. Harvey, and M. Slatkin), pp. 187–205. Cambridge University Press.

Gittleman, J. L. (1986). Carnivore life history patterns: allometric, phylogenetic and ecological associations. *American Naturalist* **127,** 744–71.

Gittleman, J. L. and Harvey, P. H. (1982). Carnivore home-range size, metabolic needs and ecology. *Behavioral Ecology and Sociobiology* **10,** 57–64.

Gittleman, J. L. and Kot, M. (1991). Adaptation: statistics and a null model for estimating phylogenetic effects. *Systematic Zoology*.

Givnish, T. J. (1980). Ecological constraints on the evolution of breeding systems in seed plants: dioecy and dispersal in gymnosperms. *Evolution* **34,** 959–72.

Godfray, H. C. J. (1987). The evolution of clutch size in parasitic wasps. *American Naturalist* **129,** 221–33.

Golenberg, E. M., Giannasi, D. E., Clegg, M. T., Smiley, C. J., Durbin, M., Henderson, D. and Zurawski G. (1990). Chloroplast DNA sequence from a Miocene *Magnolia* species. *Nature* **344,** 656–8.

Gould, S.J. (1966). Allometry and size in ontogeny and phylogeny. *Biological Reviews* **41,** 587–640.

Gould, S. J. (1975). Allometry in primates, with emphasis on scaling and the evolution of the brain. In *Approaches to primate paleobiology* (ed. F. Szalay), pp. 244–92 Karger, Basel.

Gould, S. J. (1977). *Ontogeny and phylogeny.* Harvard University Press, Cambridge, Mass.

Gould, S. J. and Lewontin, R. C. (1979). The spandrels of San Marco and the Panglossian paradigm: a critique of the adaptationist programme. *Proceedings of the Royal Society of London, Series B* **205,** 581–98.

Gould, S. J. and Vrba, E. (1982). Exaptation—a missing term in the science of form. *Paleobiology* **8,** 4–15.

Grafen, A. (1989). The phylogenetic regression. *Philosophical Transactions of the Royal Society of London* **326,** 119–56.

Grant, P. R. (1986). *Ecology and evolution of Darwin's finches.* Princeton University Press, New Jersey.

Graves, G. R. and Gotelli, N. J. (1983). Neotropical land-bridge avifaunas: new approaches to null hypotheses in biogeography. *Oikos* **41,** 322–33.

Guilford, T. (1985). Is kin selection involved in the evolution of warning coloration? *Oikos* **45,** 31–6.

Hamilton, W. D. (1967). Extraordinary sex ratios. *Science* **156,** 477–88.

Hamilton, W. D. (1979). Wingless and fighting males in fig wasps and other insects. In *Reproductive competition and sexual selection in insects* (ed. M. S. Blum and N. A. Blum), pp. 167–220, Academic Press, New York.

Harcourt, A. H., Harvey, P. H., Larson, S. G., and Short, R. V. (1981). Testis weight, body weight and breeding system in primates. *Nature* **293,** 55–7.

Harman, H. H. (1967). *Modern factor analysis.* University of Chicago Press.

Harrison, R. G. (1989). Animal mitochondrial DNA as a genetic marker in population and evolutionary biology. *Trends in Ecology and Evolution* **4,** 6–11.

Hartigan, J. A. (1973). Minimum mutation fits to a given tree. *Biometrics* **29,** 53–65.

Harvey, P. H. (1982). On rethinking allometry. *Journal of Theoretical Biology* **95,** 37–41.

Harvey, P. H. (1985). Intrademic group selection and the sex ratio. In *Behavioural ecology: ecological consequences of adaptive behaviour* (ed. R. M. Sibly and R. H. Smith), pp. 59–73, Blackwell, Oxford.

Harvey, P. H. (1986). Issues of analysis. *Nature* **321,** 573.

Harvey, P. H. (1991). Comparing uncertain relationships: the Swedes in revolt. *Trends in Ecology and Evolution* **6.**

Harvey, P.H. (in press). *The sixtieth James Arthur lecture on the evolution of the human brain*. American Museum of Natural History, New York.

Harvey, P. H. and Bennett, P. M. (1983). Brain size, energetics, ecology and life history patterns. *Nature* **306,** 244–92.

Harvey, P. H. and Clutton-Brock, T. H. (1981). Primate home-range size, metabolic needs and ecology. *Behavioral Ecology and Sociobiology* **8,** 151–5.

Harvey, P. H. and Clutton-Brock, T. H. (1983). Allometry and evolution: the survival of the theory. *New Scientist* **98,** 312–5.

Harvey, P. H. and Clutton-Brock, T. H. (1985). Life history variation in primates. *Evolution* **39,** 559–81.

Harvey, P. H. and Harcourt, A. H. (1984). Sperm competition, testes size, and breeding system in primates. In *Sperm competition and the evolution of animal mating systems* (ed. R. L. Smith), pp. 589–600, Academic Press, New York.

Harvey, P. H. and Krebs, J. R. (1990). Comparing brains. *Science* **249,** 140–6.

Harvey, P. H. and Mace, G. M. (1982). Comparisons between taxa and adaptive trends: problems of methodology. In *Current problems in sociobiology* (ed. King's College Sociobiology Group), pp. 343–61. Cambridge University Press.

Harvey, P. H. and May, R. M. (1989). Out for the sperm count. *Nature* **337,** 508–9.

Harvey, P.H. and Pagel, M.D. (1988). The allometric approach to species differences in brain size. *Human Evolution* **3,** 461–72.

Harvey, P. H. and Pagel M D. (1989). Comparative studies in evolutionary ecology: using the database. In *Toward a more exact ecology* (ed. P. J. Grubb and J. Whittaker), pp. 209–30, Blackwell, Oxford.

Harvey, P. H. and Partridge, L. (1987). Murderous mandibles and black holes in hymenopteran wasps. *Nature* **326,** 128–9.

Harvey, P. H. and Paxton, R. J. (1981). The evolution of aposematic coloration. *Oikos* **37,** 391–3.

Harvey, P.H. and Ralls, K. (1986). Do animals avoid incest? *Nature* **320,** 575–6.

Harvey, P. H. and Read, A. F. (1988). How and why do mammalian life histories vary? In *Evolution of life histories: pattern and process from mammals* (ed. M. S. Boyce), pp. 213–32, Yale University Press, New Haven, Connecticut.

Harvey, P. H. and Zammuto, R. M. (1985). Patterns of mortality and age at first reproduction in natural populations of mammals. *Nature* **315,** 319–20.

Harvey, P. H., Kavanagh, M. and Clutton-Brock, T. H. (1978). Canine tooth size in female primates. *Nature* **276,** 817–8.

Harvey, P. H., Bull, J. J., Pemberton, M., and Paxton, R. J. (1982). The evolution of aposematic coloration: a family model. *American Naturalist* **119,** 710–9.

Harvey, P. H., Colwell, J. W., Silvertown, J. W., and May, R. M. (1983). Null models in ecology. *Annual Reviews of Ecology and Systematics* **14,** 189–211.

Harvey, P. H., Martin, R. D. and Clutton-Brock, T. H. (1987). Primate life histories in comparative perspective. In *Primate societies* (ed. B. B. Smuts, D. L. Cheney, R. M. Seyfarth, R. W. Wrangham, and T. T. Struhsaker), pp. 181–96. University of Chicago Press.

Harvey, P. H., Pagel, M. D., and Rees, J. A. (1990). Mammalian metabolism and life histories. *American Naturalist.*

Harvey, P. H., Read, A. F., and Promislow, D. E. L. (1989). Life history variation in placental mammals: unifying the data with the theory. *Oxford Surveys in Evolutionary Biology* **6,** 13–31.

Harvey, P. H., Read, A. F., John, J. L., Gregory, R. D., and Keymer, A. E. (1991). An evolutionary perspective. In *Parasitism: coexistence or conflict? ecological, physiological and immunological aspects* (ed. A. Aeschlimann and C. A. Toft), pp. 344–55. Oxford University Press.

Heise, D. R. (1975). *Causal analysis.* Wiley, New York.

Hendy, M.D. and Penny, D. (1989). A framework for the quantitative study of evolutionary trees. *Systematic Zoology* **38,** 297–309.

Hennig, W. (1950). *Grundzüge einer theorie der phylogenetischen systematik.* Deutscher Zentralverlag, Berlin.

Hennig, W. (1965). Phylogenetic systematics. *Annual Review of Entomology* **10,** 97–116.

Hennig, W. (1966). *Phylogenetic systematics.* University of Illinois Press, Urbana.

Herre, E. A. (1985). Sex ratio adjustment in fig wasps. *Science* **228,** 896–8.

Herre, E. A. (1987). Optimality, plasticity and selective regime in fig wasp sex ratios. *Nature* **329,** 627–9.

Hinde R. A. (1975). The concept of function. In *Function and evolution of behaviour* (ed. G. Baerends, C. Beer, and A. Manning), pp. 3–15, Clarendon Press, Oxford.

Hixson, J. E. and Brown, W. M. (1986). A comparison of the small ribosomal RNA genes from the mitochondrial DNA of the great apes and humans: sequence, structure, evolution, and phylogenetic implications. *Molecular Biology and Evolution* **3,** 1–18.

Hofman, M. A. (1983*a*). Energy metabolism, brain size and longevity in mammals. *Quarterly Review of Biology* **58,** 495–512.

Hofman, M. A. (1983*b*). Evolution of the brain in neonatal and adult placental mammals: a theoretical approach. *Journal of Theoretical Biology* **105,** 317–22.

Höglund, J. (1989). Size and plumage dimorphism in lek-breeding birds: a comparative analysis. *American Naturalist* **134,** 72–87.

Holmes, E.B. (1980). Reconsideration of some systematic concepts and terms. *Evolutionary Theory* **5,** 35–87.

Huey, R. B. (1987). Phylogeny, history and the comparative method. In *New directions in ecological physiology* (ed. M. E. Feder, A. F. Bennett, W. Burggren, and R. B. Huey), pp. 76–198, Cambridge University Press.

Huey, R. B. and Bennett, A. F. (1986). A comparative approach to field and laboratory studies in evolutionary biology. In *Predator-prey relationships: perspectives and approaches for the study of lower vertebrates* (ed. M. E. Feder and G. V. Lauder), pp. 82–96, University of Chicago Press.

Huey, R. B. and Bennett, A. F. (1987). Phylogenetic studies of co-adaptation: preferred temperatures versus optimal performance temperatures of lizards. *Evolution* **41,** 1098–115.

Hull, D. L. (1988). *Science as a process*. University of Chicago Press.

Hutchinson, G. E. and MacArthur, R. H. (1959). A theoretical ecological model of size distributions among species of animals. *American Naturalist* **93,** 117–25.

Hutchinson, J. (1969). *Evolution and phylogeny of flowering plants*. Academic Press, New York.

Huxley, J. S. (1932). *Problems of relative growth*. Methuen, London.

Imbrie, J. (1956). Biometrical methods in the study of invertebrate fossils. *Bulletin of the American Museum of Natural History* **108,** 217–52.

Inger, R. F. (1967). The development of phylogeny of frogs. *Evolution* **21,** 369–84.

Iwabe, N., Kuma, K., Hasegawa, M., Osawa, S., and Miyata, T. (1989). Evolutionary relationship of archaebacteria, eubacteria, and eukaryotes inferred from phylogenetic trees of duplicated genes. *Proceedings of the National Academy of Sciences U.S.A.,* **86,** 9355–9.

Jarman, P. J. (1974). The social organisation of antelope in relation to their ecology. *Behaviour* **48,** 215–67.

Jarman, P. J. (1988). On being thick-skinned: dermal shields in large mammalian herbivores. *Biological Journal of the Linnaean Society* **36,** 169–91.

Jeffreys, A. J. (1987). Highly variable minisatellites and DNA fingerprinting. *Biochemical Society Transactions* **15,** 309–17.

Jeffreys, A. J., Wilson, V., and Thein, S. L. (1985). Hypervariable "minisatellite" regions in human DNA. *Nature* **314,** 67–73.

Kempthorne, O. (1957). *An introduction to genetic statistics*. Wiley, New York.

Kleiber, M. (1932). Body size and metabolism. *Hilgardia* **6,** 315–53.

Kleiber, M. (1961). *The fire of life: an introduction to animal energetics*. Wiley, New York.

Kluge, A. G. and Farris, J. S. (1969). Quantitative phyletics and the evolution of anurans. *Systematic Zoology* **18,** 1–32.

Kluge, A. G. and Strauss, R. E. (1985). Ontogeny and systematics. *Annual Review of Ecology and Systematics* **16,** 247–68.

Kondo, R., Satta, Y, Matsura, E. T., Ishiwa, H., Takahata, N., and Chigusa, S. I. (1990). Incomplete maternal transmission of mitochondrial DNA in *Drosophila. Genetics* **126**, 657–63.

Krebs, J. R. and McCleery, R. H. (1984). Optimization in behavioural ecology. In *Behavioural ecology, an evolutionary approach* (2nd edn), (ed. J. R. Krebs and N. B. Davies), pp. 91–121, Blackwell, Oxford.

Krebs, J.R., Sherry, D.F., Healy, S.D., Perry, V.H., and Vaccarino, A.L. (1989). Hippocampal specialization in food-storing birds. *Proceedings of the National Academy of Sciences U.S.A.* **86,** 1388–1392.

Kreitman, M. (1987). Molecular population genetics. *Oxford Surveys in Evolutionary Biology* **4,** 38–60.

Krimbas, C. B. (1984). On adaptation, neo-Darwinism, tautology, and population fitness. *Evolutionary Biology* **17,** 1–57.

Kuhry, B. and Marcus, L. F. (1977). Bivariate linear models in biometry. *Systematic Zoology* **26,** 201–9.

Lack, D. (1968). *Ecological adaptations for breeding in birds*. Methuen, London.

Lande, R. (1979). Quantitative genetic analysis of multivariate evolution applied to brain:body size allometry. *Evolution* **33,** 402–16.

Lapicque, L. (1907). Le poids encéphalique en fonction du poids corporel entre individus d'une mème espèce. *Bulletin Mem. Soc. Anthropol. (Paris)* **8,** 313–45.

Larson, A. (1984). Neontological inferences of evolutionary pattern and process in the salamander family Plethodontidae. *Evolutionary Biology* **17,** 119–217.

Latter, B. D. H. (1973*a*). The estimation of genetic divergence between populations based on gene frequency data. *American Journal of Human Genetics* **25,** 247–61.

Latter, B. D. H. (1973*b*). Measures of genetic distance between individuals and populations. In *Genetic structure of populations* (ed. N. E. Morton), pp. 27–39, University of Hawaii Press, Honolulu.

Lauder, G. V. (1981). Form and function: structural analysis in evolutionary morphology. *Paleobiology* **7,** 430–42.

Le Gros Clark, W. E. and Sonntag, C. F. (1926). A monograph of *Orycteropus afer* III. The skull. *Proceedings of the General Meeting of Scientific Business, Zoological Society of London* 445–85.

Le Masurier, A. D. (1987). A comparative study of the relationship between host size and brood size in *Apanteles* spp. (Hymenoptera: Braconidae). *Ecological Entomology* **12,** 383–98.

Le Quesne, W. J. (1969). A method of selection of characters in numerical taxonomy. *Systematic Zoology* **18,** 201–5.

Leutenegger, W. and Kelley, J. T. (1977). Relationship of sexual dimorphism in canine size and body size to social, behavioral and ecological correlates in anthropoid primates. *Primates* **18,** 117–36.

Levinton, J. (1988). *Genetics, paleontology and macroevolution.* Cambridge University Press.

Lewontin, R. C. (1974). *The genetic basis of evolutionary change.* Columbia University Press, New York.

Lewontin, R. C. (1978). Adaptation. *Scientific American* **239,** 156–69.

Lewontin, R. C. (1979). Sociobiology as an adaptationist program. *Behavioral Science* **24,** 5–14.

Lewontin, R. C. (1982). *Human diversity. Scientific American*, New York.

Li, H., Gyllensten, U. B., Cui, X., Saiki, R. K., Erlich, H. A. and Arnheim, N. (1988). Amplification and analysis of DNA sequences in a single human sperm and diploid cells. *Nature* **335,** 414–7.

Liao, H., McKenzie, T., and Hageman, R. (1986). Isolation of a thermostable enzyme variant by cloning and selection in a thermophile. *Proceedings of the National Academy of Sciences (U.S.A.)* **83,** 576–80.

Liem, K. F. (1973). Evolutionary strategies and morphological innovations: cichlid pharyngeal jaws. *Systematic Zoology* **22,** 425–41.

Liem, K. F. (1980). Adaptive significance of intra- and interspecific differences in the feeding repertoires of cichlid fishes. *American Zoologist* **20,** 295–314.

Liem, K. F. and Wake, D. B. (1985). Morphology: current approaches and concepts. In *Functional vertebrate morphology* (ed. M. H. Hildebrand, D. M. Bramble, K. F. Liem, and D. B. Wake), pp. 366–377, Harvard University Press, Cambridge, Mass.

Lindstedt, S.L. (1985). Birds. In *Non-mammalian models for research in aging* (ed. F. A. Lints), pp. 1–21, Karger, Basel.

Lindstedt, S. L. and Swain, S. D. (1988). Body size as a constraint of design and function. In *Evolution of life histories of mammals: theory and pattern* (ed. M. S. Boyce), pp. 93–106, Yale University Press, New Haven, Connecticut.

Lindstedt, S. L., Miller, B. J., and Buskirk, S. W. (1986). Home range, time and body size in mammals. *Ecology* **67,** 413–8.

Losos, J.B. (1990). Concordant evolution of locomotor behaviour, display rate and morphology in *Anolis* lizards. *Animal Behaviour* **39,** 879–890.

Lunneborg, C.E. (1985). Estimating the correlation coefficient: the bootstrap approach. *Psychological Bulletin* **98,** 209–15.

Lynch, M. (in prep.). Methods for the analysis of comparative data in evolutionary ecology.

MacArthur, R.H. and Wilson, E.O. (1967). *The theory of island biogeography*. Princeton University Press, New Jersey.

Mace, G. M. and Harvey, P. H. (1983). Energetic constraints on home range size. *American Naturalist* **121,** 120–32.

Maddison, W. P. (1989). Reconstructing character evolution on polytomous cladograms. *Cladistics* **5,** 365–77.

Maddison, W. P. (1990). A method for testing the correlated evolution of two binary characters: are gains or losses concentrated on certain branches of a phylogenetic tree. *Evolution* **44,** 539–57.

Maddison, W. P. and Maddison, D. R. (1989). Interactive analysis of phylogeny and character evolution using the computer program MacClade. *Folia Primatologica* **53,** 190–202.

Maddison, W. P., Donoghue, M. J., and Maddison, D. R. (1984). Outgroup analysis and parsimony. *Systematic Zoology* 33, 83–103.

Martin, R. D. (1981). Relative brain size and basal metabolic rate in terrestrial vertebrates. *Nature* **293,** 57–60.

Martin, R. D. (1983). *Human brain evolution in an ecological context*. American Museum of Natural History, New York.

Martin, R. D. and Harvey, P. H. (1985). Brain size allometry: ontogeny and phylogeny. In *Size and scaling in primate biology* (ed. W. L. Jungers), pp. 147–73. Plenum, New York.

Martins, E. P. and Garland, T. H. (1991) Phylogenetic analyses of the correlated evolution of continuous characters: a simulation study. *Evolution*.

May, R. M. (1978). The dynamics and diversity of insect faunas In *Diversity of insect faunas* (ed. L.A. Mound and N. Waloff), pp. 188–204, Symposia of the Royal Entomological Society, Volume 9.

Maynard Smith, J. (1978). Optimization theory in evolution. *Annual Review of Ecology and Systematics* **9,** 31–56.

Maynard Smith J. (1980). A new theory of sexual investment. *Behavioral Ecology and Sociobiology* **7,** 247–51.

Maynard Smith, J. and Holliday, R. (1979). Preface. In *The evolution of adaptation by natural selection* (ed. J. Maynard Smith and R. Holliday), pp. v-vii. The Royal Society, London.

Mayr, E. (1969). *Principles of systematic zoology*. McGraw-Hill, New York.

Mayr, E. (1982). Adaptation and selection. *Biologisches Zentralblatt* **101,** 161–74.

McMahon, G., Davis, E., and Wogan, G. N. (1987). Characterisation of c-Ki-*ras* oncogene alleles by direct sequencing of enzymatically amplified DNA from carcinogen-induced tumours. *Proceedings of the National Academy of Sciences (U.S.A.)* **84,** 4974–8.

McMahon, T. A. (1983). *Muscles, reflexes and locomotion.* Princeton University Press, New Jersey.

McMahon, T. A. and Bonner, J. T. (1983). *On size and life.* Scientific American, New York.

McNab, B. K. (1963). Bioenergetics and the determination of home range size. *American Naturalist* **97,** 133–40.

McNab, B. K. (1980). Food habits, energetics and population biology of mammals. *American Naturalist* **116,** 106–24.

McNab, B. K. (1986*a*). Food habits, energetics and reproduction of marsupials. *Journal of Zoology* **208,** 595–614.

McNab, B. K. (1986*b*). The influence of food habits on the energetics of eutherian mammals. *Ecological Monographs* **56,** 1–19.

Meacham, C. A. and Estabrook, G. F. (1985). Compatibility methods in systematics. *Annual Review of Ecology and Systematics* **16,** 431–46.

Millar, J. S. (1977). Adaptive features of mammalian reproduction. *Evolution* **31,** 370–86.

Millar, J. S. and Zammuto, R. M. (1983). Life histories of mammals: an analysis of life tables. *Ecology* **64,** 631–5.

Milton, K. and May, M. L. (1976). Body weight, diet and home range area in primates. *Nature* **259,** 459–62.

Mitchell, P. C. (1901). On the intestinal tract of birds, with remarks on the valuation and nomenclature of zoological characters. *Transactions of the Linnean Society of London (Zoological Series 2)* **8,** 173–275.

Miyazaki, J. M. and Mickevich, M. F. (1982). Evolution of *Chesapecten* (Mollusca: Bivalvia, Miocene-Pliocene) and the biogenic law. *Evolutionary Biology* **15,** 369–409.

Moore, G. W., Barnabas, J. and Goodman, M. (1973). A method for constructing maximum parsimony ancestral amino acid sequences on a given network. *Journal of Theoretical Biology* **38,** 459–85.

Møller, A. P. (1988*a*). Ejaculate quality, testes size and sperm competition in primates. *Journal of Human Evolution* **17,** 479–88.

Møller, A. P. (1988*b*). Testes size, ejaculate quality and sperm competition in birds. *Biological Journal of the Linnaean Society* **33,** 273–83.

Møller, A. P. (1989). Ejaculate quality, testes size and sperm production in mammals. *Functional Ecology* **3,** 91–6.

Moritz, C., Dowling, T. E., and Brown, W. M. (1987). Evolution of mitochondrial DNA: relevance for population biology and systematics. *Annual Review of Ecology and Systematics* **18,** 269–92.

Nei, M. (1976). Mathematical models of speciation and genetic distance. In *Population genetics and ecology* (ed. S. Karlin and E. Nevo), pp. 723–65. Academic Press, New York.

Nelson, G. (1979). Cladistic analysis and synthesis: principles and definitions, with a historical note on Adamson's *Familles des plantes* (1763–1764). *Systematic Zoology* **28,** 1–21.

Nelson, G. and Platnick, N. (1981). *Systematics and biogeography: cladistics and vicariance*. Columbia University Press, New York.

Nelson, G., and Platnick, N. I. (1984). Systematics and evolution. In *Beyond neo-Darwinism* (ed. M. W. Ho and P. J. Saunders), pp. 143–58, Academic Press, New York.

Newell, N. D. (1949). Phyletic size increase, an important trend illustrated by fossil invertebrates. *Evolution* **3,** 103–24.

Orzack, S. H. (1986). Sex-ratio control in a parasitic wasp, *Nasonia vitripennis*. II. Experimental analysis of an optimal sex-ratio model. *Evolution* **40,** 341–56.

Oster, G. F. and Wilson, E. O. (1978). *Caste and ecology in the social insects*. Princeton University Press, New Jersey.

Packer, C. (1983). Sexual dimorphism: the horns of African antelopes. *Science* **221,** 1191–3.

Pagel, M. D. and Harvey, P. H. (1988*a*). Recent developments in the analysis of comparative data. *Quarterly Review of Biology* **63,** 413–40.

Pagel, M. D. and Harvey, P. H. (1988*b*). How mammals produce large-brained offspring. *Evolution* **42,** 948–57.

Pagel, M. D. and Harvey, P. H. (1988*c*). The taxon level problem in mammalian brain size evolution: facts and artifacts. *American Naturalist* **132,** 344–59.

Pagel, M. D. and Harvey, P. H. (1989*a*). Taxonomic differences in the scaling of brain on body size among mammals. *Science* **244,** 1589–93.

Pagel, M.D. and Harvey, P.H. (1989*b*). Comparative methods for examining adaptation depend on evolutionary models. *Folia Primatologica* **53,** 203–20.

Pagel, M. D. and Harvey, P. H. (1990). Diversity in the brain size of newborn mammals: allometry, energetics or life history tactics. *Bioscience* **40,** 116–122.

Pagel, M. D., May, R. M. and Collie, A. (1991). Ecological aspects of the geographic distribution and diversity of mammal species. *American Naturalist*.

Partridge, L. and Harvey, P. H. (1988). The ecological context of life history evolution. *Science* **241,** 1449–55.

Patterson, C. (1980). Cladistics. *Biologist* **27,** 234–40.

Paul, C. R. C. (1982). The adequacy of the fossil record. In *Problems of phylogenetic reconstruction* (ed. K. A. Joysey and A. E. Friday), pp. 75–117. Academic Press, London.

Peters, R. H. (1983). *The ecological implications of body size.* Cambridge University Press, Cambridge.

Pickering, J. (1980). Larval competition and brood sex ratios in the gregarious parasitoid *Pachysomoides stupidus. Nature* **283,** 291–92.

Promislow, D. E. L. and Harvey, P. H. (1990). Living fast and dying young: a comparative analysis of life history variation among mammals. *Journal of Zoology* **220,** 417–437.

Quinn, T. W. and White, B. N. (1987). Identification of restriction-fragment-length polymorphisms in genomic DNA of the Lesser Snow Goose (*Anser caerulescens caerulescens*). *Molecular Biology and Evolution* **4,** 126–43.

Quiring, D. P. (1941). The scale of being according to the power formula. *Growth* **5,** 301–27.

Ralls, K. and Harvey, P. H. (1985). Geographic variation in size and sexual dimorphism of North American weasels. *Biological Journal of the Linnaean Society* **25,** 119–67.

Ralls, K., Harvey, P.H. and Lyles, A.M. (1986). Inbreeding in natural populations of birds and mammals. In *Conservation biology: the science of scarcity and diversity* (ed. M.E. Soulé), pp. 35–56. Sinauer, Sunderland, Mass.

Rayner, J. M. V. (1985). Linear relations in biomechanics: the statistics of scaling functions. *Journal of Zoology* **206,** 415–39.

Read, A. F. (1987). Comparative evidence supports the Hamilton and Zuk hypothesis on parasites and sexual selection. *Nature* **327,** 68–70.

Read, A.F. (1989). Comparative analyses of reproductive tactics. D. Phil. thesis, University of Oxford.

Read, A. F. and Harvey, P. H. (1989). Life history differences among the eutherian radiations. *Journal of Zoology* **219,** 329–53.

Reynolds, J., Weir, B. S., and Cockerham, C. C. (1983). Estimation of the coancestry coefficient: basis for a short-term genetic distance. *Genetics* **105,** 767–79.

Ridley, M. (1983*a*). *The explanation of organic diversity: the comparative method and adaptations for mating.* Oxford University Press.

Ridley, M. (1983*b*). Can classification do without evolution? *New Scientist* **100,** 647–51.

Ridley, M. (1986*a*). *Evolution and classification: the reformation of cladism.* Longman, London.

Ridley, M. (1986*b*). The number of males in a primate troop. *Animal Behaviour* **34,** 1848–58.

Ridley, M. (1988). Mating frequency and fecundity in insects. *Biological Reviews* **63,** 509–49.

Riska, B. (1991). Evolutionary models in evolutionary allometry. *American Naturalist.*

Riska, B. and Atchley, W. R. (1985). Genetics of growth predict patterns of brain size evolution. *Science* **229,** 668–71.

Robertson, J. S., Bootman, J. S., Newman, R., Oxford, J. S., Daniels, R.D., Webster, R.G. and Schild, G.C. (1987). Structural changes in the haemagglutinin which accompany egg adaptation of an influenza A(H1N1) virus. *Virology* **160,** 31–7.

Romero-Herrera, A.E., Lehmann, H., Joysey, K.A., and Friday, A.E. (1978). On the evolution of myoglobin. *Philosophical Transactions of the Royal Society of London* **283,** 61–163.

Root, R. B. (1967). The niche exploitation pattern of the blue-gray gnatcatcher. *Ecological Monographs* **37,** 317–50.

Rothschild, M. (1972). Secondary plant substances and warning coloration in insects. In *Insect/plant relationships* (ed. H. F. van Emden), pp. 59–83, Blackwell, Oxford.

Sæther, B. E. (1988). Evolutionary adjustment of reproductive traits to survival rates in European birds. *Nature* **331,** 616–7.

Saiki, R. K., Scharf, S., Faloona, F., Mullis, K. B., Horn, G. T., Erlich, H. A. and Arnheim, N. (1985). Enzymatic amplification of globin genomic sequences and restriction site analysis for diagnosis of sickle cell anaemia. *Science* **230,** 1350–4.

Sandell, M. (1989). Ecological energetics, optimal body size and sexual dimorphism: a model applied to the stoat, *Mustela erminea* L. *Functional Ecology* **3,** 315–24.

Sankoff, D. and Cedergren, R. J. (1983). Simultaneous comparisons of three or more sequences related by a tree. In *Time warps, string edits, and macromolecules: theory and practice of sequence comparison* (ed. D. Sankoff and J. B. Kruskal), pp. 253–63, Addison-Wesley, Reading, Mass.

Sankoff, D. and Rousseau. P. (1975). Locating the vertices of a Steiner tree in an arbitrary metric space. *Mathematical Programming* **9,** 240–6.

Sarich, V. M. (1977). Rates, sample sizes, and the neutrality hypothesis for electrophoresis in evolutionary studies. *Nature* **265,** 24–8.

Sarich, V.M., Schmid, C.W., and Marks, J.M. (1989). DNA hybridization as a guide to phylogenies: a critical analysis. *Cladistics* **5,** 3–32.

Schaeffer, B., Hecht, M. K., and Eldredge, N. (1972). Phylogeny and paleontology. *Evolutionary Biology* **6,** 31–46.

Schluter, D. (1984). Morphological and phylogenetic relations among the Darwin's finches. *Evolution* **38,** 921–30.

Schmidt-Nielsen, K. (1984). *Scaling: why is animal size so important?* Cambridge University Press, Cambridge.

Schoener, T. W. (1968). Sizes of feeding territories among birds. *Ecology* **49,** 704–26.

Schoener, T. W. (1983). Simple models of optimal feeding-territory size: a reconciliation. *American Naturalist* **121,** 608–29.

Seim, E. and Sæther, B. E. (1983). On rethinking allometry: which regression model to use. *Journal of Theoretical Biology* **104,** 161–8.

Selander, R. K. (1972). Sexual selection and dimorphism in birds. In *Sexual selection and the descent of man* (ed. B. Campbell), pp. 180–230, Heinemann, London.

Sessions, S.K. and Larson, A. (1987). Developmental correlates of genome size in Plethodontid salamanders and their implications for genome evolution. *Evolution* **41,** 1239–51.

Sherry, D.F., Vaccarino, A.L., Buckenham, K., and Herz, R.S. (1989). The hippocampal complex of food-storing birds. *Brain, Behavior and Evolution* **34,** 308–317.

Shine, R. (1978). Sexual size dimorphism and male combat in snakes. *Oecologia* **33,** 269–77.

Shine, R. (1979). Sexual selection and sexual dimorphism in the amphibia. *Copeia* **2,** 297–306.

Short, R. V. (1977). Sexual selection and the descent of man. In *Proceedings of the Canberra symposium on reproduction and evolution*, pp. 3–19, Australian Academy of Sciences, Canberra.

Short, R. V. (1979). Sexual selection and its component parts, somatic and genital selection, as illustrated by man and the great apes. *Advances in the Study of Behaviour* **9,** 131–58.

Short, R. V. (1981). Sexual selection in man and the great apes. In *Reproductive biology of the great apes* (ed. C. E. Graham), pp. 319–41, Academic Press, New York.

Sibley, C. G. and Ahlquist, J. E. (1983). Phylogeny and classification of birds based on the data of DNA-DNA hybridization. *Current Ornithology* **1,** 245–92.

Sibley, C. G. and Ahlquist, J. E. (1984). The phylogeny of the hominoid primates as indicated by DNA-DNA hybridization. *Journal of Molecular Evolution* **20,** 2–15.

Sibley, C. G. and Ahlquist, J. E. (1985). The phylogeny and classification of passerine birds, based on comparisons of the genetic material, DNA. In *Proceedings XVIII congressus internationalis ornitholigicus*, ACTA (ed. Ilychev, V. D. and V. M. Gavrilov), pp. 83–121. Nauka, Moscow.

Sibley, C. G. and Ahlquist, J. E. (1987). Avian phylogeny reconstructed from comparisons of the genetic material, DNA. In *Molecules and morphology in evolution: conflict or compromise* (ed. C. Patterson), pp. 95–122. Cambridge University Press.

Sibley, C.G., Ahlquist, J.E., and Monroe, B.L. (1988). A classification of the living birds of the world based on DNA-DNA hybridization studies. *Auk* **105,** 409–23.

Sillén-Tullberg, B. (1988). Evolution of gregariousness in aposematic butterfly larvae: a phylogenetic analysis. *Evolution* **42,** 293–305.

Simms, D. A. (1979). North American weasels: resource utilization and distribution. *Canadian Journal of Zoology* **57,** 504–20.

Simpson, G. G. (1945). The principles of classification and a classification of the mammals. *Bulletin of the American Museum of Natural History* **85,** i-xvi, 1–350.

Simpson, G.G. (1961). *Principles of animal taxonomy.* Columbia University Press, New York.

Simpson, G. G. (1967). *The meaning of evolution.* Yale University Press, New Haven, Connecticut.

Simpson, G. G. (1978). Variations and details of macroevolution. *Paleobiology* **4,** 217–21.

Smith, R. J. (1980). Rethinking allometry. *Journal of Theoretical Biology* **87,** 97–111.

Sneath, P. H. A. and Sokal, R. R. (1973). *Numerical taxonomy.* W. H. Freeman, San Francisco.

Snell, O. (1891). Das Gewicht des Gehirns und des Hirnmantels der Säugetiere in Beziehung zu deren geistigen Fähigkeiten. *Sitzungsberichte Ges. Morph. Physiol. (Munchen)* **7,** 90–4.

Sober, E. (1989). *Reconstructing the past: parsimony, evolution, and inference.* Massachusetts Institute of Technology Press, Cambridge, Mass.

Sokal, R. R. and Rohlf, F. J. (1969). *Biometry* (1st edn). Freeman, New York.

Sokal, R. R. and Rohlf, F. J. (1981). *Biometry* (2nd edn). Freeman, New York.

Sokal, R. R. and Sneath, P. H. A. (1963). *Principles of numerical taxonomy.* W. H. Freeman, San Francisco.

Southwood, T. R. E. (1961). The number of species of insect associated with various trees. *Journal of Animal Ecology* **30,** 1–8.

Sprent, P. (1969). *Models in regression and related topics.* Methuen, London.

Springer, M. and Krajewski, C. (1989). DNA hybridization in animal taxonomy: a critique from first principles. *Quarterly Review of Biology* **64,** 291–318.

Stearns, S. C. (1983). The influence of size and phylogeny on patterns of covariation among life-history traits in mammals. *Oikos* **41,** 173–87.

Stearns, S.C. (1984). The tension between adaptation and constraint in the evolution of reproductive patterns. *Advances in Invertebrate Reproduction* **3,** 387–398.

Stephens, D. W. and Krebs, J. R. (1986). *Foraging theory*. Princeton University Press, New Jersey.

Stevens, P. F. (1980). Evolutionary polarity of character states. *Annual Review of Ecology and Systematics* **11,** 333–58.

Stewart, C.-B., Schilling, J.W., and Wilson, A.C. (1987). Adaptive evolution in the stomach lysozymes of foregut fermenters. *Nature* **330,** 401–404.

Stone, G. N. and Willmer, P. G. (1989). Warm-up rates and body temperature in bees: the importance of body size, thermal regime and phylogeny *Journal of Experimental Biology* **147,** 303–328.

Straney, D. O. and Patton, J. L. (1980). Phylogenetic and environmental determinants of geographic determinants of geographic variation of the pocket mouse *Perognathus goldmani* Osgood. *Evolution* **34,** 888–903.

Sugihara, G. (1980). Minimal community structure: an explanation of species abundance patterns. *American Naturalist* **116,** 770–787.

Sutherland, W. J., Grafen, A., and Harvey, P. H. (1986). Life history correlations and demography. *Nature* **320,** 88.

Swofford, D. L. and Berlocher, S.H. (1987). Inferring evolutionary trees from gene frequency data under the principle of maximum parsimony. *Systematic Zoology* **36,** 293–325.

Swofford, D. L. and Maddison, W. P. (1987). Reconstructing ancestral character states under Wagner parsimony. *Mathematical Bioscience* **87,** 199–229.

Taylor, C. R. and Weibel, E. R. (1981). Design of the mammalian respiratory system. 1. Problem and strategy. *Respiratory Physiology* **44,** 1–10.

Taylor, P. D. (1981). Intra-sex and inter-sex sibling interactions as sex determinants. *Nature* **291,** 64–66.

Templeton, A. R. (1987). Genetic systems and evolutionary rates. In *Rates of Evolution* (ed. K. S. W. Campbell and M. F. Day), pp. 218–234. Australian Academy of Sciences, Canberra.

Terborgh, J. and Goldizen, A.W. (1985). On the mating system of the cooperatively breeding saddle-backed tamarin. *Behavioral Ecology and Sociobiology* **16,** 293–9.

Thompson D'Arcy W. (1917). *On growth and form.* Cambridge University Press.

Thorpe, J. P. (1982). The molecular clock hypothesis. *Annual Review of Ecology and Systematics* **13,** 139–68.

Toro, M. (1981). Sex ratio variation and evolution. D. Phil. thesis. University of Sussex, Brighton.

Toro, M. and Charlesworth, B. (1982). An attempt to detect genetic variation in sex ratio of *Drosophila melanogaster*. *Heredity* **49,** 199–209.

Trevelyan, R., Harvey, P. H. and Pagel, M. D. (1990). Metabolic rates and life histories in birds. *Functional Ecology*, **4,** 135–141.

van Berkum, F. H. (1986). Evolutionary patterns of the thermal sensitivity of sprint speeds in *Anolis* lizards. *Evolution* **40,** 594–604.

Van Valen, L. (1973). Body size and numbers of plants and animals. *Evolution* **27,** 27–35.

Vawter, L. and Brown, W. M. (1986). Nuclear and mitochondrial DNA comparisons reveal extreme rate variation in the molecular clock. *Science* **234,** 194–6.

von Baer, K. E. (1828). *Ueber Entwickelungsgeschichte der Thiere. Beobachtung und Reflexion*. Borntrager, Königsberg.

Waage, J. K. (1982). Sib-mating and sex-ratio strategies in Scelionid wasps. *Ecological Entomology* **7,** 103–12.

Walker, I. (1967). Effect of population density on the viability and fecundity in *Nasonia vitripennis* Walker (Hymenoptera, Pteromalidae). *Ecology* **48,** 294–301.

Wayne, R.K. and O'Brien, S.J. (1987). Allozyme divergence within the Canidae. *Systematic Zoology* **36,** 339–355.

Wayne, R.K., Benveniste, R.E., Janczewski, D.N. and O'Brien, S.J. (1989). Molecular and biochemical evolution of the Carnivora. In *Carnivore behavior, ecology, and evolution* (ed. J.L. Gittleman), pp. 465–95, Comstock Associates, Cornell University Press, Ithaca.

Weibel, E. R. and Taylor, C. R. (1981). Design of the mammalian respiratory system. *Respiratory Physiology* **44,** 1–164.

Werren, J. H. (1980). Studies in the evolution of sex ratios. Ph.D.thesis, University of Utah. Salt Lake City.

Werren, J. H. (1983). Sex ratio evolution under local mate competition in a parasitic wasp. *Evolution* **37,** 116–24.

West Eberhard, M. J. (1978). Polygyny and the evolution of social behavior in wasps. *Journal of the Kansas Entomological Society* **51,** 832–56.

Western, D. and Ssemakula, J. (1982). Life history patterns in birds and mammals and their evolutionary interpretation. *Oecologia* **54,** 281–90.

Wiley, E. O. (1975). Karl R. Popper, systematics and classification: a reply to Walter Bock and other evolutionary systematists. *Systematic Zoology* **24,** 233–43.

Williams, G. C. (1966). *Adaptation and natural selection*. Princeton University Press, New Jersey.

Wilson, A. C., Carlson, S. S., and White, T. J. (1977). Biochemical evolution. *Annual Reviews of Biochemistry* **46,** 573–639.

Wilson, A. C., Cann, R. L., Carr, S. M., George, M., Gyllensten, U. B., Helm-Bychowski, K. M., Higuchi, R. G., Palumbi, S. R., Prager, E. M., Sage, R. D. and Stoneking, M. (1985). Mitochondrial DNA and two perspectives on evolutionary genetics. *Biological Journal of the Linnaean Society* **26,** 375–400.

Winterbottom, J. M. (1929). Studies in sexual phenomena. VI. Communal display in birds. *Proceedings of the Zoological Society of London* 186–95.

Wölters, J. and Erdmann, V. A. (1986). Cladistic analysis of 58 rRNA and 168 rRNA secondary and primary structure—the evolution of eukaryotes and their relation to Archaebacteria. *Journal of Molecular Evolution* **244,** 152–66.

Wooton, J. T. (1987). The effects of body mass, phylogeny, habitat, and trophic level on mammalian age at first reproduction. *Evolution* **41,** 732–49.

Wu, C. F. J. (1986). Jacknife, bootstrap and other resampling methods in regression analysis. *The Annals of Statistics,* **14,** 1261–95.

Wright, S. (1932). *The roles of mutation, inbreeding, crossbreeding, and selection in evolution.* Proceedings of the XI International Congress of Genetics **1,** 356–66.

Wyles, J. and Gorman, G. C. (1980). The albumin immunological and Nei electrophoretic distance correlation: a calibration for the saurian genus *Anolis* (Iguanidae). *Copeia* **3,** 66–71.

Author Index

Subject Index